Actuaries' Survival Guide

Actuaries' Survival Guide

Navigating the Exams as Applications of Data Science

Third Edition

Ping Q. Wang, PhD, ASA, CERA
Professor
Greenberg School of Risk Management,
Insurance and Actuarial Science
St John's University, New York
New York
United States

Fred E. Szabo, PhD
Professor
Department of Mathematics and Statistics
Concordia University
Montreal
Canada

ELSEVIER

Academic Press is an imprint of Elsevier
125 London Wall, London EC2Y 5AS, United Kingdom
525 B Street, Suite 1650, San Diego, CA 92101, United States
50 Hampshire Street, 5th Floor, Cambridge, MA 02139, United States

Notices

Knowledge and best practices in this field are constantly changing. As new
research and experience broaden our understanding, changes in research methods,
professional practices, or medical treatment may become necessary.

Practitioners and researchers must always rely on their own experience and
knowledge in evaluating and using any information, methods, compounds, or
experiments described herein. In using such information or methods they should be
mindful of their own safety and the safety of others, including parties for whom they
have a professional responsibility.

To the fullest extent of the law, neither the Publisher nor the authors, contributors, or
editors, assume any liability for any injury and/or damage to persons or property as
a matter of products liability, negligence or otherwise, or from any use or operation
of any methods, products, instructions, or ideas contained in the material herein.

ISBN: 978-0-443-15497-3

For Information on all Academic Press publications
visit our website at https://www.elsevier.com/books-and-journals

Publisher: Peter B. Linsley
Senior Editorial Project Manager: Sara Valentino
Publishing Services Manager: Shereen Jameel
Project Manager: Vishnu T. Jiji
Senior Designer: Christian J. Bilbow

Typeset by MPS Limited, Chennai, India

Printed in India

Last digit is the print number: 9   8   7   6   5   4   3   2   1

Working together
to grow libraries in
developing countries

www.elsevier.com • www.bookaid.org

"Prediction is very difficult, especially about the future."
Niels Bohr, Physicist and Nobel Laureate
"It is better to be roughly right than precisely wrong."
Carveth Read, Philosopher and Logician

Contents

Acknowledgments

The third edition of the *Actuaries' Survival Guide* is an adopted brainchild to me. Katey Birtcher, then Publisher of STEM Education Content, now Publishing Director at Elsevier Academic Press, contacted me in July 2021, inquiring about my interest in revising a book on the actuarial profession and education intended for students and young professionals.

Dr. Fred E. Szabo profiled the activities, career paths, and requirements of actuaries in the first two editions of *Actuaries' Survival Guide*. I found the book informative and comprehensive, indeed a valuable manual for both students and young professionals.

Wishing to keep the legacy and usefulness of the book, and knowing that Dr. Szabo was not available to revise the third edition, I decided to take on the challenge.

I want to thank both Dr. Szabo and Ms. Birtcher for giving me an opportunity to reach out to the actuarial education community beyond St John's University, my employer.

Preparatory work kicked off in the spring of 2022. Thanks to the help of Jody Queen-Hubert, then Director of Ellen Thrower Center for Apprenticeship and Career Services, Greenberg School of Risk Management, St John's University, I compiled a contact list of about 200 actuarial alumni who had all taken one or more of my classes on various subjects: probability and statistics, financial mathematics, loss modeling, and life contingencies. A questionnaire based on Dr. Szabo's original one, complemented with new items, was sent to my former students, now practitioner actuaries in every branch of actuarial practices. Their responses to the survey constitute an indispensable component of the third edition.

In the era of digital information, the Internet has become the first source of reference. However, I quickly realized that the richness of information does not guarantee correctness due to various reasons. Language barrier was another obstacle when I was trying to collect data on actuarial organization and examination systems from countries across the world, even with the help of Google Translate. Individuals (most are related to the actuarial organization of each country) in the following list helped me in verifying the information I collected about their country's actuarial education and examination system, or in directly providing such information:

Argentina: Javier Campelo (through Andres Kopyto)
Belgium: Rob De Staelen
China: Yuantao Xie, Xiufang Li
Denmark: Frederikke Horn

Finland: Petri Martikainen
France: Samuel Cywie
Germany: Martin Brandt, Mariella Linkert
Hong Kong: Joanna Cheung
India: Swetha Jain
Ireland: Paul Williams
Italy: Giampaolo Crenca
Japan: Shoko Matsui
Netherlands: Leandra Pennartz
Norway: Hans Michael Øvergaard
Portugal: Lourdes Afonso
South Africa: Justina Komana
Spain: Instituto de Actuarios Españoles
Switzerland: Holger Walz

Elizabeth Smith of Casualty Actuarial Society, Aleshia Zionce, Toni Ugarte, and Brett Rogers of Society of Actuarial Society helped me secure permission to use sample questions and answers of CAS and SOA examinations and other information.

Printed hard-copy materials still serve as a solid and reliable source of intelligence. When I need such material, Ismael Rivera-Sierra, Director of Kathryn and Shelby Cullom Davis Library of St John's University, proves, as always, the best friend of a researcher.

To all those whose help has made this edition possible, I hereby express my sincere gratitude.

A special thanks is in order for Sara Valentino, Senior Editorial Project Manager, and Vishnu T. Jiji, the project manager for this edition. Their timely and gracious reminders kept me on track with this project. Finally, the Electronic Manuscript Submission System made it so easy to submit all the work and keep track of all revisions. I would like to say "thank you" to the IT team who makes the process enjoyable.

Ping Q Wang
ASA, CERA
July 8, 2023

Introduction

You are reading this book because you are thinking about the future. What would you like to do with your life? What career would allow you to fulfill your dreams of success? If you like mathematics, your choices have just become simpler. Consider becoming an actuary. About 20 years ago, Dr. Fred E. Szabo wrote the first edition of the *Actuaries' Survival Guide* to explain what actuaries are, what they do, and where they do it. The second edition was published 10 years after the first one because of its remarkable popularity among readers.

The publisher, Elsevier Academic Press, believes that the book's legacy and usefulness should be continued. However, Dr. Szabo is not available for the task and is willing to hand the book off to someone else, so the editor in charge of the project contacted me midyear of 2021.

That was the first time that I learned about the existence of this book. I read the book twice, first as a reader and then from the perspective of a (potential) author.

My conclusions were:

1. The book is indeed valuable to two groups with strong mathematics backgrounds: students in high school or college and professionals who are seeking a second chance at a career.
2. The book needs updating to keep up with the ever-changing examinations and syllabi of actuarial education. Since the birth of the book, the examination structures and exam syllabi of both the Society of Actuaries and the Casualty Actuarial Society have undergone several rounds of restructuring and change. You can find a list in Chapter 1 that describes the effective periods of some examinations.

Changes have taken place since the second edition of this book:

- Distinct syllabi. The examination structures of the Society of Actuaries and the Casualty Actuarial Society of America have been modified multiple times. The assessment of computational skills now is part of the exam systems.
- Data science. Big data has become a popular topic since the late 2000s and the data scientist profession has attracted many college students since then. The skill sets required of data scientists and of actuaries are quite similar. Which career will you choose, actuary or data scientist?

- Test delivery system. Most examinations are now computer based, and examinees of a number of exams can know the preliminary results almost instantly after they submit their work.
- Exam exemptions. The SOA has begun to give approved college courses credits for certain exams since the fall of 2022. The CAS recognizes the exam waivers granted by the Canadian Institute of Actuaries University Accreditation Program.
- Online tutoring. As in many other fields, the internet has changed the way in which actuarial students study and interact. In addition to taking actuarial courses at college, students turn to online tutoring platforms for more help. Some platforms offer video instructions along with written answers and solutions. The flexibility in the design and delivery of the vendors' products even makes it possible for instructors to incorporate the online exercises and assessments as required components of students' coursework.

Social networking and global expansion, as noted by Dr. Szabo, continue to be important topics and phenomena of the actuarial community.

I decided to take the challenge. Being an associate of the Society of Actuaries and a Chartered Enterprise Risk Analyst, I stay at the forefront of the actuarial profession. I have taught college courses in the subjects of probability, financial mathematics, loss modeling, and life contingencies, both in the United States and abroad for about a quarter of a century; I connect with former students and stay informed of changes in employers' requirements of entry-level actuaries.

Despite these changes in exam structure and syllabi, the principles, skills, and techniques on which the actuarial profession is based and depends have remained stable. A survey of my former students was conducted and the responses to the survey were very similar to those that appeared in the first edition.

Since the courses and examinations required to become an accredited actuary are subject to ongoing changes, references to courses and examinations in this book are meant as illustrations only. You should always check with the societies' websites for up-to-date course syllabi and all matters related to actual examinations.

I sincerely hope this edition will continue the legacy and usefulness of this work, started by Dr. Szabo, for students and early professionals.

What Is an Actuary?

You may have heard several descriptions of actuaries and their work. Are they mathematicians, statisticians, economists, investment bankers, legal experts, accountants, business experts, or data scientists? This book will show you that an actuarial career involves elements of all of these professions and more. One goal of this book is to open the rich mosaic of actuarial life for you. As such, this book is expected to be more than a career guide; rather, a career companion on the road to your professional success.

The Syllabi

As you explore the different features of this guide, you will appreciate its global perspective. You are contemplating becoming an actuary. It therefore makes sense to consider all of your options so you can decide for yourself why you might want to do so and assess how to succeed in one of the most desirable professions.

You have probably heard that you need to study long and hard to become an accredited actuary. But what do you need to study and how does the subject you are required to master in North America, for example, compare with that required elsewhere? By comparing different syllabi you will quickly get an idea of what is involved. You can use this book to compare several actuarial syllabi:

1. The syllabi of the International Actuarial Association
2. The syllabi of the Actuarial Association of Europe
3. The syllabi of the Society of Actuaries
4. The syllabi of the Casualty Actuarial Society

When you look at the sample examination questions and their solutions, you are of course not meant to know it all. You are encouraged, however, to relate some of the topics to your complementary course and get a sense of the academic direction your studies should take.

In addition, you can use the "Actuaries around the World" section to compare these syllabi with the accreditation systems in several other countries and, by perusing the reciprocity agreements between different countries described in Appendix C, you can develop a feel for the remarkable international portability of actuarial credentials, unmatched by almost any other profession.

The international nature of the actuarial profession is again illustrated in Appendices A and B, where the national and international locations of many of the listed companies are included to allow you to dream of working in some of your favorite cities, states, provinces, and countries around the globe.

You may also wonder why the basic actuarial symbols are illustrated in this guide. One obvious reason is that some of them are used in the sample examination questions in the book. But a deeper reason is the universality of the actuarial language, which adds another dimension to the portability of actuarial knowledge, experience, and techniques across national borders. Once again we recognize actuaries as a global community of experts.

The Examinations

Another useful feature of the book is the inclusion of sample questions from the Society of Actuaries (SOA) and Casualty Actuarial Society (CAS) examinations.

By browsing through these sections, you will get an idea of what the examination questions look like and get a feel for what the answers should be. Although the form and content of these examinations will continue to change over time, the ideas and techniques presented in the given examples will give you an idea of the mindset of professional actuaries.

Chances are that neither the questions nor their answers will make sense to you at this time. But by perusing their content and looking at the form of their answers, you will get a sense of what lies ahead.

In order to pass the SOA and CAS examinations, you must use some of the techniques and study tools discussed in Chapter 2. You will find a list of appropriate references on the websites listed in Appendix D.

Global Employers

A further aspect of this book that you will find useful is a list of typical employers. The list is incomplete because there are thousands of public and private companies as well as government agencies employing actuaries. As presented, the list is meant as a starting point for your personal research into actuarial employment. The details provided about different companies differ from employer to employer. The intention is to highlight different aspects of employment rather than giving an encyclopedic description of employment at particular companies. You can easily complete the sketches by consulting the cited websites. They typify the extraordinary opportunities that exist worldwide for actuarial employment.

Online Resources

Several online resources have been created to highlight the fact that the actuarial profession is essentially a close-knit family of professionals that cares deeply about the success of its members. This is probably one of the reasons why ongoing surveys have shown actuaries belong to "one of the most desirable professions."

The website entitled Be An Actuary (see www.beanactuary.org) is definitely required reading for aspiring actuaries at all levels: high school, college, and university. The site contains answers for students, parents, educators, counselors, and employers.

In addition, online resources such as the Actuarial Bookstore (see www.actuarial-bookstore.com) have publications with information on various topics, including how to prepare for your first actuarial examination, beginning your actuarial career, overviews of skills and exams, actuarial internships, full-time positions, and employer actuarial student programs.

On the website of the Society of Actuaries you can find a section called Future Actuaries. It is an excellent source of information on what actuaries do, actuarial credentials, actuarial education and scholarships, and career development including internship opportunities.

Resources in Print

Much of the material integrated in this book is in the public domain and is available in bits and pieces through a multiplicity of published sources. However, it is widely scattered and incoherent and, as such, appears disjointed and overwhelmingly complex. One of the objectives of this project was to analyze the available information and build a coherent picture.

The sources of the quoted material are always clear from the context and are often explicitly mentioned. When required, italics are used to identify passages quoted verbatim. However, the world does not stand still and the source of the quoted material may have changed over time. This is especially important to keep in mind in the case of actuarial examinations and the rules that govern them. You should always consult the up-to-date websites or latest editions of published material for revisions of content, rules, and requirements. The same applies to things such as reciprocity agreements and other legal matters. The guide is primarily meant to be a motivating roadmap through the maze of the actuarial profession, enriched with enlightening, constructive, and supporting comments from working actuaries, actuarial employers, and actuarial students.

Chapter 1

Actuarial careers

1.1 Career Options

The word *actuary* comes from the Latin word *actuarius*, which referred to shorthand writers in the days when things like typewriters and computers hadn't even been dreamed of. Today, actuaries work for insurance companies, consulting firms, government departments, financial institutions, and other organizations like Amazon and Uber. They provide crucial predictive data analysis upon which major business decisions are based. True to their historical roots, actuaries still use a rather extensive shorthand for many of the special mathematical functions required for the work (see Bowers et al. (1997), pages 687–691; Perryman (1949), pages 123–131; Appendix H). The sample questions and answers presented in Chapter 2 illustrate some of the currently used actuarial symbols listed in Appendix H. The symbols are an amazingly rich combination of right and left subscripts and superscripts, attached to designated upper- and lowercase Roman and Greek letters. They provide a unique and common actuarial language around the world.

Actuarial science is an exciting, always-changing profession, based on fields such as mathematics, probability and statistics, economics, finance, law, and business. Most actuarial work requires knowledge and understanding of all of these fields and more. To ensure that this is really the case, all actuaries must pass special examinations before being recognized as members of the profession. To perform their duties effectively, actuaries must also keep abreast of economic and social trends, as well as stay up to date on legislation governing areas such as finance, business, health care, and insurance.

No doubt you have heard about the actuarial examinations you need to pass to become an Associate or Fellow of one of the actuarial societies. Often full-time employees in actuarial firms who are still engaged in the examination-writing process are distinguished from Associates and Fellows by being referred to as Student Actuaries. The efforts required to succeed in these examinations are in many ways analogous to those required to become a doctor, lawyer, or other high-ranking professional. So are the rewards. Citing findings by *US News and World Report*, *CNN Money*, and the *Jobs Rated Almanac*

Actuaries' Survival Guide, Third Edition. DOI: 10.1016/B978-0-443-15497-3.00001-2

(see Kranz and Lee (2015)), a website jointly sponsored by both the Casualty Actuarial Society and the Society of Actuaries, www.beanactuary.org, claims that "no matter the source, actuary is consistently rated as one of the best jobs in America." In its online publication *Occupational Outlook Handbook*, the Bureau of Labor Statistics, a branch of the United States Department of Labor, announces that the job outlook for actuaries from 2021 to 2031 is 21%, a pace much faster than the average of about 5%.

Actuaries are experts in the assessment and management of risk. Traditionally, the risks managed by them have been insurance and pension funding risks, although the management of business risks is also among the responsibilities of insurance actuaries. So is the insurance of insurance, known as reinsurance. Moreover, many actuaries are now also managing asset-related risks in merchant banks and consulting firms. This bodes well for the long-term future of the profession since risks of all kinds will always be with us. The global financial crises plaguing the modern world highlight the importance of being able to measure financial risks and being able to develop plans for managing them. However, as you will see later on in this book, the day-to-day activities of an actuary depend very much on the sector of the financial services industry where the actuary works.

Actuaries are often chosen to be chief executive officers in insurance companies. This is because actuaries possess deep insight into the risk and profitability of the company and upper management and boards of directors have a high regard for the knowledge and skills of actuaries, and because the need of a company to maintain its financial integrity makes an actuary's numerical skills invaluable.

Actuarial terms, acronyms, and definitions

As you read on, you will quickly discover that actuarial science is full of technical terms, acronyms, and definitions. This book is not the place for explaining them in detail, because the definitions involved are readily available in textbooks and on the Internet. The main objective of this book is to introduce you to the career opportunities that exist in the actuarial world and to sketch for you the steps required to enter that world. For this reason, most of the technical material in the book is provided only in illustrative and summary form. Consider it a detailed roadmap to the relevant topics in mathematics, statistics, and business. It is merely meant to help you identify the range of knowledge involved in actuarial work. The study of the mentioned topics requires specialized sources and tools. The reference section at the end of the book provides you with the necessary pointers.

Actuaries can be grouped in different ways. As their functions change in response to changes in the world around us, the distinctions become less sharp. However, the following categories of employment will give you an initial idea.

Valuation actuaries

Reserves are important to the long-term financial health of an insurance company. Because insurance companies are dealing with events that are uncertain in time and amount, they must put aside what they consider to be the most likely amount of money (reserve is the technical term) they will need to pay for future claims and expenses, and then put aside a little more, just in case. The role of valuation actuaries is to determine the appropriate "just a little more" and to validate the expected number of claims and amounts to be taken into account when setting the price of insurance. Valuation actuaries also certify required reserves to insurance regulators.

Pricing actuaries

Pricing actuaries are responsible for determining how much money a company is likely to charge on a product. A product can be life insurance that pays an agreed-upon sum to your beneficiary when you, the insured person, die; an annuity that pays an agreed-upon sum every month as long as you live; or some form of health insurance that covers the costs of medical care not paid for by a government plan, for example, dental and drug expenses. A product can be homeowner insurance that pays, upon the occurrence of covered damage to your home, an amount determined according to the severity of the damage and the amount of coverage purchased. A product can also be automobile insurance that pays for the damage to the vehicle or covers the driver's financial liability arising from accidents. Pricing actuaries use the same assumptions as valuation actuaries when calculating the price of insurance to guarantee consistency and ensure that when valuation actuaries believe that they are adding a little extra to the reserves, they are really doing so. We might say that pricing actuaries add pricing recipes to insurance products to make them profitable.

Valuation actuaries and pricing actuaries work mainly for insurance companies including life insurance companies, property and casualty insurance companies, health insurance companies, and reinsurance companies.

Pension actuaries

Pension actuaries look at the ages and salaries of all members of a pension plan and project how much benefit each would receive at retirement on average, given consideration of various future events, including that some members will terminate before retirement, and some will get salary increases. Then they look at the assets in which the pension plan has invested and determine, based on these two analyses, how much the plan's sponsor (usually an employer) needs to contribute to the plan each year. The pension actuary certifies that the contributions needed to fund the plan are adequate and qualify for a tax deduction for the sponsor. Pension laws and pension regulations are country specific or even state specific (United States) or province specific, as is the case in Canada. This is the one area in which the global mobility of actuaries is somewhat restricted.

Special examinations must be passed in the country of employment to be a pension actuary.

In the United States, an actuary must be an Enrolled Actuary to have signing authority to perform actuarial services required under the Employee Retirement Income Security Act of 1974 (ERISA). Pension actuaries must be Enrolled Actuaries to be eligible to perform government-related pension fund audits. Enrolled actuaries are also employed in the human resource departments of large companies. Senior pension actuaries in the United States are usually also Fellows of the American Society of Pension Professionals and Actuaries (ASPPA), a designation that is awarded only after successful completion of a series of professional examinations. The basic examinations are those required to become an Enrolled Actuary, together with three additional ASPPA examinations. A Fellow of the American Society of Pension Professionals and Actuaries must also be a Fellow or Associate of one of the following societies, namely the Society of Actuaries (SOA), the Casualty Actuarial Society (CAS), the Canadian Institute of Actuaries, or the Institute and Faculty of Actuaries, or be a member of the American Academy of Actuaries (AAA), the Mexican Association of Consulting Actuaries, the Mexican Association of Actuaries, or the National College of Actuaries of Mexico.

The Employee Retirement Income Security Act of 1974 specifies that they must have participated in determining that the methods and assumptions adopted in the procedures and followed in actuarial services are appropriate in the light of all pertinent circumstances. They must also demonstrate a thorough understanding of the principles and alternatives involved in such actuarial services. Their actuarial experience must include involvement in the valuation of the liabilities of pension plans, wherein the performance of such valuations requires the application of principles of life contingencies and compound interest in the determination, under one or more standard actuarial cost methods, of such of the following as may be appropriate in the particular case: normal cost, accrued liability, payment required to amortize a liability or other amount over a period of time, and actuarial gain or loss.

In the United Kingdom, Canada, and certain other countries, Appointed Actuaries play a role analogous to that of Enrolled Actuaries in the United States.

Consulting actuaries

Consulting actuaries advise clients on insurance, investment, and pension-related decisions. They complete risk and cost analysis and determine where financial uncertainties lie using their knowledge in statistics, economics, and laws and regulations. Consulting actuaries who specialize in pensions spend a good deal of their time advising on clients' benefit programs and defined benefit pension plans. These are trusts set up to fund tax-assisted retirement benefits at a rate spelled out in a legally certified document. Thanks to the access to data and expertise in all types of actuarial work, consulting actuaries also provide actuarial services to insurance companies when the carrier's own in-house actuaries need help.

In the United States, senior consulting actuaries are usually members of the Conference of Consulting Actuaries (CCA). They must be a Fellow or Associate of the Society of Actuaries or the Casualty Actuarial Society; or be enrolled with the Joint Board for the Enrollment of Actuaries (EA), thus having acquired the title of Enrolled Actuary; or be a member of the American Academy of Actuaries (AAA); or be a member of the American Society of Pension Professionals and Actuaries (ASPPA).

Finance/investment actuaries

Actuaries can find that their skills are needed in fields of capital management, corporate finance, investment, and mergers and acquisitions. More and more, financial actuaries tend to become "risk actuaries," employed to model types of financial risks. An advertisement for a senior financial actuary on the Internet describes one of the novel roles of actuaries in business. A company was looking for a senior financial actuary whose responsibilities included "*developing, analyzing, and testing models of Internet credit card processing systems including product pricing, positioning, and consumer credit, in order to minimize risk and improve return on investment. You will communicate assumptions, results, and alternatives to staff and provide guidance in systems reengineering.*" A suitable candidate was expected to have at least a Bachelor's degree in actuarial science, finance, mathematics, or a related field and be an Associate Actuary. In addition to appropriate experience, the candidate was expected to be an effective communicator, and creative thinking skills were essential. The company was looking for a self-starter with a strong statistical background and proven expertise in modeling techniques. Moreover, knowledge of the financial and management needs of an Internet real-time credit card processing company was expected.

Enterprise risk management actuaries

In June 2017, the Institute and Faculty of Actuaries in the UK published a paper entitled "Risk Management—An Actuarial Approach" (see IFoA (2017)). In its preamble, the paper states that Actuaries are skilled professionals whose comprehensive training includes the use of statistical analysis to understand risks and uncertainties. They are therefore well placed to support organizations' risk management efforts. An actuarial approach to risk management places a particular focus on measuring and understanding the impact of risks, both positive and negative, on the outcomes experienced and considering how the risks and their impacts may evolve over time. The paper is predated by another from the same organization in 2011, "Actuaries in Risk Management" (see IFoA (2011)). The Institute and Faculty of Actuaries believes that "*with their deep expertise in risk, actuaries are widening their influence with careers moving beyond financial services into newer areas such as healthcare and energy. The actuarial qualification provides a foundation that is both broad and technically detailed, helping to equip actuaries to play to their personal strengths to expand their universe of opportunities. It is, therefore, no surprise to find actuaries aiming to be to the fore in the emerging fields of risk management, including the development of risk-based approach to managing an enterprise."*

Since the mid-1990s, enterprise risk management (ERM) has emerged as a concept and as a management function within corporations. Enterprise risk management is a systematic and integrated approach to the management of the total risks that a company faces. ERM is now a key component of the syllabi of both the SOA and the CAS. Recognizing the unique and advantageous role of actuaries in this realm, the SOA created the Chartered Enterprise Risk Analyst (CERA) designation in 2007. Now the Global Association of CERA administers the CERA credentials worldwide and receives recognition from actuarial professional bodies around the globe. The CERA designation of actuaries has become an enticing career opportunity for actuaries around the world.

What does it take to become an actuary?

Skills needed to become an actuary certainly include mathematical ability, knowledge of and comfort with computers and computer modeling systems, data analysis, and the ability to communicate complex topics in terms that customers, oftentimes nonactuaries, can understand.

Formally, most actuarial positions require that you are at least an Associate of the Society of Actuaries, the Casualty Actuarial Society, or the Canadian Institute of Actuaries, or have equivalent standing in an actuarial society of another country. If you are in a position that requires you to certify actuarial valuations and reports, you must usually be a Fellow of one of the professional societies.

Many actuaries in the United States are also members of the American Academy of Actuaries (see Appendix D), the public policy, communications, and professionalism organization for all actuaries in the United States.

As Section 1.14 shows, actuaries in different countries belong to a wide variety of national and international professional organizations that define and direct the future of the profession. At the international level, the International Association of Actuaries (see Appendix D) plays a central role in coordinating and advancing global actuarial interests.

Rapid and profound socioeconomic changes around the globe bring with them specific problems that need to be solved and financial risks that need to be understood and managed by actuaries. Situations arise that require the actuaries attending to these problems be exceptionally creative and able "to think outside the box." The growing emphasis on on-the-job experience and continued professional development by actuarial societies reflects this change.

1.2   Benefits and Rewards

During Dr. Wang's two dozen years of teaching actuarial science as professor, associate of SOA, and a CERA, several hundred students have been in his classes and many of them remain active on his LinkedIn contact list. He surveyed them for this edition and found the responses were in general similar to the findings reported in the previous two editions by Dr. Szabo, except for those on the latest technology developments. These surveyed actuaries all have at least one thing in common—they all love mathematics, although some end up doing nonactuarial work. Here is what some of them have given as reasons for their career choice.

Q: **Did you ever consider working in a nonactuarial field of applied mathematics (such as engineering)? If so, what tipped the scales in favor of an actuarial career?**

More than 50% of all respondents to the survey said, "No." There was no doubt in their minds that all they ever wanted to be was an actuary. Some actually switched to actuarial career from other professions, like mathematics teacher or biology major. Here is what some of them had to say.

ANS: No, I've never considered a nonactuarial career.

ANS: I am now a software engineer. Compensation tipped the scales for software engineering.

ANS: I was a biology major prior to my actuarial masters. I think the actuarial program and career provides many benchmarks which I appreciated. The test process allows you to track your success.

ANS: I have and still am. The majority of opportunities presented to me have been actuarial roles to this point.

ANS: Yes, and I have switched to Physics. I find the work much more rewarding. I am currently in the Medical Physics community and do research and engineering in radiation oncology.

ANS: Yes, I was a high school teacher before having a career change to become an actuary because I wanted to work in the industry where I can use my mathematical, and analytic skills and knowledge to solve complexities of the evolving world that we live in.

ANS: Not really. It's very painful to study for exams, but the career is very stable and needed.

ANS: I have not.

ANS: Not really. I find applied mathematics too technical and actuarial science has a good combination of math/economics/finance.

ANS: I did not consider any other major. I pursued actuarial science because I wanted a strong foundation for my actuarial career.

ANS: I knew I didn't want to be an engineer—I was too interested in business and finance and found the actuarial exam structure very appealing. I thought about being an accountant but did not want to follow in my sibling's footsteps.

ANS: I have not.

Q: Do you know any people who switched from other professions to actuarial careers? What were their previous professions? What did it take for a successful switch to actuarial career?

ANS: Yes, from various previous positions. Music, underwriting, and accounting are a few examples.

ANS: Yes, mathematical teachers. They did successfully switch to actuarial careers.

ANS: Actuarial career is a high paying career compared to others such as teaching. To successfully switch, you need to pass exams.

ANS: A lot of people switch to actuarial careers. They didn't like the old jobs or there are no future there. If you really want to switch to actuarial, go to school and choose an actuarial/statistics major, pass some exams, and most importantly, get an actuarial internship. Without any internship, you probably won't get any interviews at all.

ANS: I am a good example.

ANS: Not very common due to the exams.

ANS: Many of the senior actuaries I talked to experienced career changes in their paths. Some of them were previously traders, engineers, etc., and they only learned about the actuarial profession midway and decided to pursue a career in it. Personally I think it takes a lot of determination, perseverance, and support from the family to switch to an actual career as they need to balance life, work, and study on their way of taking actuarial exams.

ANS: I do not know of anyone who has switched into an actuarial career.

ANS: I know of people from engineering and legal backgrounds who switch to an actuarial profession successfully. They are generally curious about learning new concepts, and love to apply mathematical principles in business.

ANS: I've heard of many individuals that were previously teachers who decided to pursue actuarial careers. I've also heard of some people who worked in insurance in other orles, such as underwriting or finance, that made the switch. These individuals tend to see the same barriers to entry as "typical" actuaries—pass the exams, deal with the entry-level "grunt" work, and work hard to get where you want to be. I think the actuarial profession is actually fairly welcoming in that sense—it does not forbid anyone from pursuing the career, so long as they're willing to put the time and effort into passing the exams.

ANS: I know someone who was an engineer and switched to actuary. It took the person a couple of months to pass two exams to be qualified enough to get a job.

1.3   Actuary vs Data Scientist

The term Data Science was first coined in 2008, and in less than one decade Data Scientist has become one of the hottest and most trending professions. Data scientists and actuaries share many common skill requirements, and both are excellent career options for students with strong mathematical backgrounds and interest. How do the two compare? Here is what actuaries had to say in the survey.

Q : What is the difference between actuaries and data scientists?

ANS: Actuaries distinguish themselves in being able to better understand the business impacts of their work and being able to communicate their work to others.

ANS: Actuaries are part data scientists since we deal with large data sets but with more skills.

ANS: Actuaries need more industry knowledge and insurance specific knowledge, and pricing and reserving knowledge.

ANS: Actuaries are focused on the insurance industry. Both professions analyze data and draw conclusions, but data scientists usually use more programming skills to do so.

ANS: Actuaries do both jobs in my opinion. Data scientists only do the data manipulation and cleaning.

ANS: Actuaries use mathematics, statistics, data analysis, financial and accounting knowledge and skills to minimize the cost of risk exposure.

Data scientists are professionals who use machine learning to solve complexities of issues involving datasets.

ANS: Both actuaries and data scientists are supporting underwriting teams. Pricing actuaries need to understand products and help underwriters improve the performance and achieve their business goals; reserving actuaries work much closer with claim teams to set reserve/sensitivities and estimate ultimate losses. From my understanding, both pricing and reserving actuaries need to know something regarding products/ business units. Data scientists work with data to build models, they don't need to know a lot regarding products or the company's performance, but they know the whole industry since they can collect the industry data and build models with it. For example, data scientists build a model with industry data to grade their insureds based on their financial performance. Underwriters can use the score to decide if they want to keep this insured or not; if they keep this insured for one policy term, actuaries can do some product analysis or client analysis by calculating ultimate loss ratio to show if the insured is worth keeping.

ANS: Actuaries need to know actuarial concepts and apply them to lculating ultimate loss ratio to show business practices.

ANS: Data scientists focus purely on analyzing data. While that is part of the actuarial profession, the scope of an actuary's job is wider.

ANS: I think actuaries, especially research actuaries, and data scientists have a lot in common in that both professions are capable of dealing with a large quantity of data and have great analytical skills. However, actuaries possess more actuarial knowledge that are specific to the insurance industry while data scientists might not be as insightful on such matters. On the other hand, data scientists are more technical and possess better programming skills.

ANS: I am not sure.

ANS: In my opinion, there isn't much difference if both are successful in helping companies analyze their business other than the fact that data scientists aren't credentialed.

ANS: Actuaries have a much deeper affiliation with the insurance industry while data scientists theoretically take any data and do an analysis on them, but not fully understanding the business implications.

ANS: Actuaries are a lot more knowledgeable of business and finance than data scientists. Data scientists are very technical and analytical, but their expertise typically ends when it comes to drawing conclusions or communicating results to management. There does tend to be quite a bit of overlap between the two, but generally data scientists know when to take a step back and say, "I did my part with giving the actuaries the information/tools, now it's up to them to draw the conclusions and take the results to management."

ANS: Actuaries communicate outputs of projects to managers/executives. Data scientists know how to program certain parts of projects to get the output.

ANS: Actuaries are trained to consider other sources of information to render an opinion. Data scientists focus primarily on data driven results. Additionally, there are robust standards and professionally recognized organizations and associations that help maintain the qualification standards of actuaries.

Q: Do you think data scientists will replace actuaries in the near future? Why or why not?

ANS: I think there should be a good balance of both. I don't think having a team of solely data scientists is beneficial as they don't have as deep an understanding of insurance necessarily.

ANS: Just being able to infer trends in the data does not necessarily mean the data scientist will be able to predict accurately.

ANS: No, because one needs professional training to be an actuary. Data scientists do not have insurance industrial knowledge and actuarial insights.

ANS: No, they do not have the insurance expertise needed.

ANS: No, I do not think so, data manipulation is a small part of our job. Drawing conclusions based on actuarial knowledge is the difficult part.

ANS: No, there is no way data scientists will replace actuaries because data scientists are subsets of the actuarial industry. Actuaries use the skills in data science in their daily lives.

ANS: No, because they focus on different things. Data scientists are usually PhDs while actuaries are not. Each business unit needs actuaries for sure, but data scientists generally work based on projects/models and they don't have to limit themselves to a specific line of business.

ANS: No, because data scientists have to acquire actuarial knowledge in order to do insurance work.

ANS: Not, actuaries have wider scope jobs, and we are learning data science techniques to apply these methods to our work.

ANS: I don't think so. Actuaries possess insurance-specific knowledge that cannot be replaced by data scientists. Additionally, there has been an increasing emphasis on data analysis and programming in the actuarial curriculum. Actuaries are capable of what data scientists do but with insurance insights.

ANS: I do not think so as they are two different disciplines.

ANS: Just as it is hard to pick up R or Python and do modeling, it is equally hard to learn insurance concepts. Once a student is in the actuarial profession, as long as they are keeping up with the analytical needs of insurance industry, it shouldn't be a huge concern.

ANS: No—I think data scientists are not specialized and don't have the business sense to do what an actuary can do.

ANS: No, but I think they will start to look more and more alike. I think the testing structure needs to be modernized to reflect the way an actuary practices today and that will involve a lot more skills that look like ones required in the data scientist field.

ANS: Definitely not. There are so many more functions that actuaries perform that data scientists cannot, not should they ever. Actuaries have a specific set of knowledge about insurance and probability that enables them to understand and communicate results in a way that data scientists simply aren't trained to do. Plus, regulators will always need an actuary to sign off on reserves and rates, but I don't think one will ever wholly replace the other.

ANS: I do not think so, since data scientists aren't as prepared to communicate output of results to managers/executives compared to actuaries.

ANS: No.

ANS: I believe that without expanding the actuarial toolkit, the allure of becoming an actuary will diminish, not because the skillset is replaceable; the profession needs to be able to attract talent.

1.4 A Typical Workday

You may wonder what a day in the life of an actuary is like. What are the typical tasks, and how does the day evolve? Obviously the answers depend on the nature of the company and the seniority of the actuary.

Here is what several actuaries and actuarial students had to say about this in the survey.

Q: Please describe a typical workday in the life of an actuary.

ANS: Excel is still the primary tool for actuaries. Lots of Excel. Research on a given line of business, keeping up with the industry. Researching impacts of inflation more recently. Collaborating with colleagues. Working with members of other teams (underwriters, cat modelers, clients, among others).

ANS: There are no typical days. It all depends on the project you are working on at my company especially as a consultant.

ANS: Setting up the schedule of a project, setting up meetings and doing preparation of meetings. Doing filing documentations.

ANS: Attend meetings, adjust assumptions in models, explain results, email others about business context.

ANS: A typical day in the life of an actuary starts with checking emails for any fires. Prepare monthly tasks such as IBNR (incurred but not reported) or renewal work. Reach out to my actuary if there is anything she needs help on. Present conclusions and recommendations in a formal email.

ANS: A typical day in an actuary is about cleaning dataset, validating dataset, running checks, developing loss development triangles using Excel, RStudio, doing reserve analysis using industry benchmarks, BF (Bornhuetter-Ferguson) method, RAA (Random Access Algorithm) method, writing emails and responding to emails, a team meeting and meeting deadlines.

ANS: The first thing in the morning is checking emails, if there is something needed from me, I will reply to the email and send whatever they need; I will have an informal meeting with my manager to go through several things I am working on. Spend some time with my intern, guiding them with some projects. Continue on my current projects based on priorities.

ANS: A typical day for an actuary involves managing emails in the morning—responding, delegating or adding emails to the project list. Then the actuaries will keep working on their projects or review deliverables from the team. There are meetings during the day, either with the internal team or with other departments and sometimes clients, to discuss project details, progress, encountered challenges, and resolutions. Corporate actuaries generally interact more within the company while consulting actuaries have more client-facing moments in their jobs. Unless in the busy season, actuaries normally can maintain a good work-life balance and have an exciting personal life.

ANS: Wake up, coffee and breakfast, commute to office if needed, study, work, lunch, work, commute again if needed, home, dinner, exercise, wind down for the day.

ANS: My current position relates to building scalable models for brokers to use, so think of it as a car, we build the look of the car, the engine of the car, work with craftsmen (UX/Developers) to ensure what we designed is what we wanted.

ANS: When I first get to work, I check my emails and see what came in since I logged off previously. Anything that I need to follow up on, gets marked with a flag to remind myself for later. I then go to my notes and check on the projects I'm working on—what the status is, who I need to reach out to, and what's a priority. I make a plan and prioritize my day, accounting for meetings and lunch. Usually I have anywhere from 2–5 meetings a day if I'm in the office, and only 1–2 if I'm working from home. I then grab a cup of tea and begin to work— lately I've been using R to create various reports. I typically grab lunch in our office's cafeteria around 11:30 am with one of my coworkers, then we bring our food back to our desks to eat. I'll take this time to slow down with my work and catch up with the world on my phone, or chat with the people sitting around me. After I finish eating, I'll get back to whatever I was working on. Every 2–3 weeks, there will be a lunchtime webcast hosted by an actuarial organization like SOA and AAA and I will usually eat lunch while attending that. Around 3 pm, other people in the office start losing their steam and wander around and make casual conversation—I'll either join in if I'm not too busy, or I'll put headphones on and listen to some music to help me focus. I typically leave the office around 5:30 pm. I stay a little late to time it so I can pick up my significant other from the train station on my way home.

ANS: Work 9–5, and then study an hour before and after work.

ANS: A typical day varies based on the time of the year. For example when it's time to file you'll be focused on filing, and when it's time for jumbo renewal, you'll be working on supporting underwriting with pricing and key assumptions.

1.5   Typical Projects and Responsibilities

How do beginning actuaries spend their time at work, and how do these activities change as an actuary's career advances?

Q: What are some of the typical actuarial projects on which you have worked, and what specific knowledge and skills were required? Please give some illustrative examples.

ANS: I've worked on a wide range of products from asbestos reserves to rate indications for Florida homeowners and offshore energy, to creating an underwriting pricing tool, ILS (insurance-linked securities) work, and others. Loss modeling was important for a client who was looking to increase their offering of higher limits on a given line of business. Various reserving methods (exam 5 material) were important in calculating default ELRs (estimated loss ratios) and catastrophe loads for specific lines of business. Excel used for mostly all projects.

ANS: Pension and retiree medical valuations, benefit calculations review. It involves being able to read plan documents and infer how the benefits should be valued.

ANS: I've been working on pricing and rate revision projects, and implementation of a new product. Needed skills and knowledge: pricing, regulation of filing, company products, rate manual, and policy rating.

ANS: Valuation projects, regulation changes, auditing assumption changes, recalculating reserves.

ANS: I work mostly with IBNR (incurred but not reported), which is integral to my job. It is important to give reserve recommendations monthly, to avoid volatility. This is used by the finance team in presenting the data to the street for our stock.

ANS: I have provided support in cleaning historical datasets which are currently being housed in Excel folders and files. And as more datasets are being added, the system becomes very slow to work with because the files have many formulas and links. Our reserving team is working on automating our historical dataset into SQL using the combination of RStudio and Excel. I copied, pasted, and ran checks for all historical data from inception to date. Also, I provided support to the pricing actuary team to test a tool prepared for underwriters to use when speaking to customers, or policyholders, and other insurers from the reinsurance industry.

ANS: On the pricing side, the typical project is profitability analysis, which involves many actuarial skills and knowledge like loss ratio calculation, on-leveling premium, Bornhuetter-Ferguson method, loss trend, loss development pattern, etc.

ANS: Creating actuarial tools to facilitate the implementation of product specifications in the system. Run actuarial modeling/valuation software and perform analysis based on the output.

ANS: Most recently, I've been working on converting Excel actuarial models into Python. Firstly, coding knowledge is imperative for such efforts. Secondly, I need to understand accounting and present values because the model performs financial statement projections. I've also worked on optimizing our capital and asset strategies. Having knowledge of capital markets and the types of assets that insurance companies can invest in is crucial for such a role.

ANS: Establishing routine in-force management processes—it requires me to use power query in Excel to analyze large amounts of data and streamline the process extensively. Pricing model conversion—it requires me to be familiar with the corporate actuarial pricing model as well as product features and to customize the database when necessary. Merger and Acquisition deals—it requires me to be familiar well with different kinds of advanced reinsurance structures, actuarial appraisals, innovative reserve financing solutions, etc.

ANS: Reserve Reviews—Exam 5 of CAS knowledge (triangles, making selections, justifying selections, presenting results); Industry Results (Using SNL Excel add-in, knowing what certain insurance terms meant; and creating presentations).

ANS: I am more or less working on nontraditional actuarial projects. We utilized data from the industry, combining it with our client historical loss data to generate a scalable Director and Officers model. This model required us to use our modeling skills to identify frequency and severity by exposure entered, and then using financial theory to identify how a company with more money might need less insurance than another company with less money.

ANS: In my first role, I was a valuation (or reserving) actuary. My work was very cyclical – every quarter we'd get an update to our in-force, we would update our assumptions, and we'd run models to project our cash flows. We did this on a GAAP, statutory, and market-consistent liability basis. I had a number of responsibilities throughout the quarter. I would sometimes be responsible for processing the raw in-force files into files that can be read by our modeling software. I'd also often make changes to the models, and I'd work on running the models in a stepwise fashion so we could analyze the impact of each change to the model separately. I would then summarize the results and review the numbers, checking for reasonableness and accuracy. This role required me to be very knowledgeable of our quarterly process—there were a number of steps we took, and it took a few quarters before I really started to understand how everything fit together. It also required me to be very well-organized and aware of the current status of my team-mates' work. I was one of the most technical actuaries on my team, so I often did a lot of work using automating Excel macros (VBA) and R scripts.

In my current role, I am a pricing actuary. This role is much more project-based and less cyclical. Most of the projects I've been working on lately have utilized R and SQL to pull and analyze data from our system's database. One of these reports is a COVID Claims dashboard, where I look at the financial impact of COVID on our Health business. Another report is a Loss Ratio analysis report, where I breakdown our loss ratio between various attributes (underwriting type, issue age, plan, etc.). I have also worked on a number of projects regarding our competitive position in the marketplace. Using an industry data set in R, I'll pull the premium rates for a particular state or part of the country and analyze how our premiums compare to our competitors. This requires not only a knowledge of R, but also knowledge of the marketplace—this includes knowing what plans and ages we want to focus on, and who are our biggest competitors.

ANS: Some typical projects have been selecting loss development factors, expected loss ratios given prior history, and selecting IBNR using various methods. Knowledge of credibility and trends in the underlying data is essential. Also, Excel skills are required, in terms of knowing the most efficient formulas to use as well as how to lay out the file in a simple manner.

ANS: Some projects I have worked on are forecast modeling to project future medical expenses, updating rate manuals to reflect current experience to update the rates when underwriting prices for renewals and new clients. Skills I need are listening, coding, and communicating the assumptions.

Entry-level jobs

Q: **What are the responsibilities of new employees in actuarial entry positions in your company, and what are their typical tasks and salary ranges?**

ANS: Typical starting salary may be around $65–75k now (though I'm not sure). Tasks are really getting to know the business and working on basic pricing or reserving projects. Nothing crazy, no decision-making on development factors or trend rates (that's for more senior members to decide). We are mostly looking for new members to update the exhibits, assess new data for reasonability, and give us their initial thoughts and observations.

ANS: Analyze data, develop programs, prepare client presentations. Not sure of latest salary ranges.

ANS: New employees are responsible for some simple tasks and easy work, and also participate in big projects with a mentor or manager's lead. Typical tasks include rate revisions and data analysis assignments. Salary ranges $65k–70k a year.

ANS: They help populate worksheets and models to generate results. $65–70k for new graduates.

ANS: I think our new hires make around $80,000 (dependent on tests). They are expected to start with no knowledge but be able to gain knowledge quickly.

ANS: New employees at reserving team at BHSI provide supporting roles in all areas such as quarterly reserve closings, running data checks and data validations, cleaning datasets and identifying incomplete dataset, learning about the business, preparing year end actuarial reports (e.g., converting documents to pdf format), proofreading of documents, and providing enough hands-on opportunity to learn about business partners, and many more.

ANS: Entry-level actuaries are doing things like price monitoring, data cleaning and mapping, profitability analysis, updating templates, etc. Salary ranges are $60k–70k, based on how many exams you passed.

ANS: Ad hoc reporting, supporting senior actuaries on various projects.

ANS: Entry-level associates are responsible for completing smaller tasks such as updating parts of an actuarial model, repeating with new data a process that was performed last year. I recommend DW Simpson or a similar agency for salary data.

ANS: Salary range is about $65k–80k depending on the number of exams passed. Typical tasks in the pricing team include routine processes such as preparing experience reports and meeting slides, running basic pricing models, market research, competitive analysis, etc.

ANS: Updating workbooks, presentations, initial pass of reserve reviews. $65,000.

ANS: New employees are usually expected to take over routine processes while learning various concepts for the work.

ANS: Data cleansing, mapping exposure information, documenting process, providing analyst support to projects. I would say a new actuarial student with two exams would be around $75k–80k.

ANS: New entry-level actuaries at my company tend to spend their first few months learning about the company and figuring out where they can bring the most value to the team. We don't have a formal rotation program, so actuaries are simply hired into a specific role—they learn, on the job, how to do what's required of them. Typically entry-level actuaries start at $70k–75k a year at my company, depending on the number of exams they've passed. And this number may have increased in recent years, due to inflation.

ANS: New employees are responsible for reserving for one business unit each quarter, as well as updating various exhibits in the off-quarter months. Salary range for an assistant is $60–65k.

ANS: Responsibilities for new employees are very basic and rudimentary. It is important to first give entry-level actuaries exposure on how pulling data works and the different products offered by the company. Typical tasks are monthly reports on trends, membership growth or decline, etc. Salary ranges vary by area.

Intermediate-level jobs

Q: What are the responsibilities of employees in intermediate actuarial positions in your company, and what are their typical tasks and salary ranges?

ANS: So many variables at play here. I would say salary ranges from $100k–200k (5–10 years of experience). Responsibility includes managing a small team (or maybe interns). Taking on larger, high-leveraged projects and making more decisions (i.e., development factors, ELFs, reserves, trend rates, ELRs—estimated loss ratios, etc.).

ANS: Client interaction and responses, review of work quality, and managing teams. $150k–160k.

ANS: Being in charge of a project and participating in decision making process. I'm not sure about the salary ranges.

ANS: They drive the auditing and modeling decisions. $120k–200k for seniors and managers.

ANS: I think intermediate positions offer $100–120k, based on exams. The tasks are to help and guide the younger new hires. Also, presenting to external team communications is expected.

ANS: The responsibilities of employees in the intermediate actuarial positions in my company are to support their respective managers. In BHSI, we work as a team so there is continued support among the team levels irrespective of the level of experience in the company.

ANS: Intermediate actuarial positions start by owning a bigger task and becoming responsible for a larger part of the project. They may become responsible for interacting with other stakeholders and can even delegate some tasks to entry-level actuaries.

ANS: Intermediate actuarial positions in my company include senior actuarial associates, managers, and directors. They are in charge of more advanced actuarial responsibilities, such as product development, in-force management, valuation, modeling, reinsurance management, etc. They are familiar with actuarial regulations and practices and demonstrate more experienced project-management and problem-solving skills. The salary range is around $90k to $180k, depending on roles and years of experience.

ANS: Helping with interns, being on top of your tasks, initial to second pass of reserve reviews, running meetings. $80k–100k.

ANS: Intermediate actuarial positions usually require the candidate to work fairly independently and own specific functions and projects on the team.

ANS: Evaluate the data analysis from data analyst, check the cleansing process and mapping process, and make recommendations about the risk of interest. I would say salary ranges around $100k–150k.

ANS: These actuaries tend to be fairly self-sufficient. They have strong understandings of the concepts, and they apply these concepts well to their work. They work on analysis that are then sent to senior leadership, and they help develop and train the more junior actuaries. I don't know what their salary range typically is.

ANS: Employees in intermediate actuarial positions are expected to work on 2–3 business units each quarter, as well as LDF/ELR (loss development factors/estimated loss ratio) analyses in between the different quarters. Salary ranges could be $80–110k.

ANS: Typical responsibilities vary based on the area you work in. For health care in pricing, a more intermediate position works on updating rate manuals and explaining key variances and assumptions to senior management.

Progression of responsibilities

Q: **What are typical SOA and CAS career paths and where should successful actuaries or actuarial students be at age 20, 25, 30, 35, 40, and 45 in (a) SOA and (b) CAS?**

ANS: This is up to each individual themselves to determine. Some are perfectly comfortable being individual contributors through their careers. Some are "live to work" people while others are "work to live" people. Some people fly through exams while others take a little longer. I don't believe there are concrete checkpoints actuarial students should check off by a given age. Too many factors play a role in what your career will look like.

ANS: It depends on the field they want to be in and based off of that a, choice between the SOA and CAS.

ANS: CAS career path is from actuarial analyst 1, actuarial analyst 2 to senior actuarial analyst, then to manager and associate actuary, and finally chief actuary. After 10 years in the career, a successful actuary should be an FCAS and do some management work. Then maybe chief actuary after 20 to 30 years.

ANS: Associate by 25, fellow by 30, and retired by 40 would be an optimal path.

ANS: I think in your mid-20s you should be showing a propensity to pass exams, by 30 you should be an ASA, by 35 an FSA, and starting in your late 30s you should be showing value to your company.

ANS: (a) For SOA designation, most successful actuaries or actuarial students will be at 20 to 35; and (b) for CAS designation, most successful actuaries or actuarial students will be at 40 to 45.

ANS: In the earlier age of the career (less than 10 years), you should pass all the exams. You then need to find which lines you are interested in and develop yourself in that direction.

ANS: I think this is highly subjective. Success means different things to different people. Some people rise up within the company quickly, but have narrow knowledge. Others can move around quite a bit and accumulate broader knowledge, which sets them up for higher leadership roles eventually.

ANS: (a) SOA: Age 20: start taking actuarial exams, actuarial internship experience. Age 25: start taking FSA exams, rotate in actuarial functions and decide on a field to work in. Age 30: get FSA designation, management roles, build a presence in the actuarial community. Later: get exposure in various advanced actuarial functions, such as reserving solutions, asset-liability management, and merger and acquisition, move to senior management roles in the company and more advanced roles in SOA or American Actuarial Association.

ANS: 20—studying for exams; 25—passed a couple of exams with an actuarial job; 30—FCAS; 35—managerial position; 40/45 – moving up the company ladder.

ANS: Not so sure.

ANS: I think my personal goal is to lead a team of my own in a smaller brokerage firm by 40.

ANS: I can only speak to SOA, since that's the society I am in. At age 20, most actuaries are still in college, so they should have 1–2 exams and hopefully an internship under their belt before graduating. At age 25, an actuary should be close to their ASA. If they graduated college with 4–5 exams, I would expect them to have their ASA by 25, barring any unexpected life events that could delay them. By 30, I would expect most actuaries to have their ASA, unless they got started in their career late (like a career changer). By 35–40 I would expect an actuary to have either decided to stop at their ASA or continue on to their FSA. They likely have important responsibilities at work but may not be manager-level yet. By 40–45, they are either an expert in their area, or have moved up into a leadership role. By 45+, they will either continue working their way up the corporate ladder, or they will find their spot and stick with it until they want to change roles.

ANS: Assuming someone starts taking exams in college, I would imagine that a successful actuary should be ASA by 25–26 and FSA by 30–35. For the CAS route, a successful actuary should be ACAS by 26–27 and FCAS by 30–35. Based on the company I could see a successful actuary being a manager by the time they're 30–35.

ANS: I think you can be successful in all actuarial areas.

Note on the survey

All the survey participants for the third edition graduated from the actuarial science program of St John's University Greenberg School of Risk Management,

Insurance and Actuarial Science between 2006 and 2019, most from the undergraduate program and several from the graduate program.

Some questions on the survey relate to current and earlier exams required by the SOA and the CAS. The list below presents the exams referenced and their active periods. Appendices F and G provide a complete description of the current SOA and CAS exam systems.

Exam 5 (Basic Techniques for Ratemaking and Estimating Claim Liabilities), CAS, since May 2007.

Exam 6 Regulation and Financial Reporting (country specific), CAS, since Oct. 2011.

Exam C (Construction and Evaluation of Actuarial Models), SOA, May 2007 to June 2018.

Exam P/1 (Probability), required by both societies, since February 2007.

Exam FM/2 (Financial Mathematics), required by both societies, since May 2007.

Exam IFM (Investment and Financial Markets), SOA, July 2018 to November 2022.

Exam LTAM (Long Term Actuarial Mathematics), SOA, July 2018 to Spring 2022.

Exam MAS I and II (Modern Actuarial Statistics), CAS, since May 2018.

Exam MLC (Models for Life Contingencies), SOA, May 2007 to April 2018.

Exam MFE (Models for Financial Economics), SOA, May 2007 to March 2018.

Exam PA (Predictive Analytics), SOA, since December 2018.

Exam S (Statistics and Probabilistic Models), CAS, October 2015 to October 2017.

Exam SRM (Statistics for Risk Modeling), SOA, since September 2018.

Exam STAM (Short Term Actuarial Mathematics), SOA, October 2018 to June 2022.

1.6   Mathematical Skills

Here is what the respondents to the survey had to say about the basic mathematical knowledge they require in their daily work. They also commented on the connection between theory and practice. People often wonder what links there are between the actuarial examinations and their required working knowledge of mathematics, finance, economics, and other special subjects such as risk theory, loss modeling, and stochastic methods.

Q: **What general mathematical competencies are required by an actuary? Give some examples and relate them to the SOA or CAS examinations.**

ANS: Frequency/Severity calculations, curve fitting—Heavy CAS—Exam 5 material. Probabilities—Exam P.

ANS: The annuity functions are heavily used in benefit calculations and also the methodology to come up with discount rates. So for me, the LTAM or MLC were the most useful material on a daily basis.

ANS: Actuaries are required to know calculus, statistics, and probability. They are required to pass exams like Exam P, Exam FM, Exam IFM, Exam MAS-I (CAS).

ANS: Algebra for all exams; calculus for exam P.

ANS: Financial mathematics and probability would be the biggest two. I think understanding interest over time is important for FM, MFE, LTAM, and a significant amount of the health track for FSA.

ANS: I believe that an actuary needs conceptual understanding of mathematical concepts like algebra (ratios, percentages, equations, polynomials, real and imaginary numbers, systems of equations), advanced knowledge of calculus, statistics, economics, basic accounting, basic programming skills (such as Structured Query Language—SQL, RStudio), advanced knowledge of Microsoft Excel and Word, advanced level writing and effective communication skills and problem solving skills. The ability of an actuary to translate real-life situations or problems into mathematical systems of equation by effectively identifying the known and unknown quantities or all the parameters in the problem, identifying constraints in the problem, and applying effective problem solving tools to solve for the unknown quantities. Then after solving the unknown quantities, the actuary should be able to provide a detailed explanation and demonstrate why the method used in solving the real life problems actually works.

ANS: For exams, you need to have a lot of statistical knowledge. You will learn a lot when preparing for exams, but in work, you may not touch too complicated math unless you work in the modeling team.

ANS: Probability, calculus, financial management, life contingency, math, and statistics.

ANS: Addition, subtraction, and basic mathematic concepts are used daily by actuaries to perform calculations.

ANS: Statistics, probability, time value of money, financial mathematics, etc. All areas mentioned above are covered in preliminary actuarial exams.

ANS: Algebra (for all CAS exams) and calculus (especially for the earlier CAS exams).

ANS: With most of the analytical work done in spreadsheet or statistical software, practicing actuaries mostly need to have good mathematical and logical thinking.

ANS: Through the preliminary exams, algebra/calculus is required for the majority of the exams; some relates to memorizations on recognizing the specific formula of density function $f(x)$ to identify mean/variance in much shorter formats. In the upper-level exam, geometry (Table M/L Charges for rate changes) are calculated through triangles/some type of shape.

ANS: An actuary needs a strong sense of logic and reason—oftentimes, the most logical solution is the best solution, and an actuary needs a strong sense of that to be successful in business. An actuary does not need to be a math genius—they should not be afraid of math, but they do not need to be in love with it. Math is a tool that we use—the better the relationship an actuary has with math, the more successful they will be.

ANS: At least basic calculus for the exams, but mostly logic steps that are required in math. I think understanding percentage changes for sensitive applications is probably more applicable in most cases.

ANS: Probability, aggregations, and understanding of distributions. An example of probability is expected value, which can be seen in pretty much every CAS exam. Aggregating data and understanding how similar/different it is from the underlying data can be found in mid-upper-level CAS exams. Distributions such as binomial and Poisson can be seen in Exam P as well as other exams.

ANS: Actuaries in health are required to know basic algebra. In healthcare, you'll be using algebra to solve premiums, and you'll need to know how to create line graphs to better analyze results.

Q: Why do actuaries need calculus? Please give examples and relate them to the SOA and CAS examinations.

ANS: Actuaries work with frequency and severity curves quite a bit. Finding areas under curves at various limits/retention levels for policies is important and must be done efficiently.

ANS: Calculus was needed since some of the preliminary exams like P, test your understanding of differentiation and integration.

ANS: Some of the relevant topics in calculus including basic integration techniques (substitution, integration by parts), improper integrals, and double integrals are helpful when you have to apply a probability distribution to model something. And calculus is needed in Exam MAS-1.

ANS: Actuaries often work with continuous functions which may involve calculus such as integration – exam P.

ANS: I think having a solid basis of calculus is important to understand how the exam questions are tailored. In many exams (P, C) they teach you a formula, but it is important to derive the variables for the formula. A lot of exam questions want you to manipulate variables into a format that the formula can handle.

ANS: Actuaries need calculus because calculus is the building block on which all real life scenarios are built on. For example, pricing actuaries will have to accurately estimate or forecast how much premiums a customer or policyholder or insurer will have to pay with the promise to compensate him/her of any future losses using benchmarks set by National Association of Insurance Commissioners (NAIC), historical data and knowledge of calculus to achieve accurate premium. In the same way, reserving actuaries will apply calculus skills to predict or forecast Ultimate Loss, Allocated Loss Adjusted claim loss expenses (ALAE), and paid Unallocated Loss Adjusted Expenses (Paid ULAE) which will ultimately develop claims. Also, a strong calculus background will help actuary to pass SOA exams P, FM, IFM, STAM, and any advanced SOA/CAS exams.

ANS: Calculus is only needed for exams. I don't see anything in work related to calculus.

ANS: Derivation is used in life contingency math a lot.

ANS: I cannot think of recent examples where I used calculus on the job.

ANS: My current daily responsibilities do not require calculus, but I imagine it would be useful for actuarial research and the way of thinking is critical to actuarial problem-solving.

ANS: Earlier CAS exams involve solving problems with calculus.

ANS: Actuaries or any analytics profession rarely apply calculus by hand, but the concept is crucial in understanding advanced statistical methods. For example, a good understanding of calculus helps candidates grasp concepts in CAS MAS I and MAS II exams, such as gradient descent for optimization, and also the advantages of the Hamiltonian Monte Carlo over other Bayesian modeling tools that rely solely on random walk.

ANS: Exam MAS 1 and 2 are related to creating loss distributions which are often used to identify risk using historical losses trended and developed. Using calculus to identify mean on a limited expectation basis, and to calculate EVPV (expected value of process variance) and VHM (variance of hypothetical mean), without calculus none of that would be possible, or it would not be as simple as it is now.

ANS: Actuaries need calculus for the core understanding of the topics covered by the preliminary exams. They will almost never use calculus directly in their jobs. It is, however, a building block to understanding probability and statistics, and those are key parts of an actuary's job.

ANS: I think for most preliminary exams, integration and derivatives are required, and that's why calculus is needed. Especially in exam P, and what used to be STAM. There are a lot of unique probability functions that require integration to get the cumulative amount. Besides that, understanding derivatives helps with FM, IFM, and LTAM. Mostly to understand formulas.

ANS: Actuaries need calculus for determining areas under the curve, which is what integrals determine. This can be found in Exam P, Exam 5, and Exam 8, amongst others.

ANS: I believe calculus is used more frequently in life insurance than in healthcare.

Q: **Why do actuaries need probability? Please give examples and relate them to the SOA or CAS examinations.**

ANS: Probability plays a role in both pricing and reserving. Being able to quantify a claims probability of reaching a layer is important in setting reserves, calculating loss elimination ratios, and stress testing.

ANS: We used decrements like mortality or retirement rates a lot as a retirement actuary. Also, as an aspiring actuary, we needed it in exam LTAM.

ANS: They'll calculate the probability of events occurring to determine the estimated funds needed to pay claims. Probability is tested heavily in Exam P.

ANS: The number one job of an actuary is to assess probability and risk. Understanding probability is fundamental. Exam P, LTAM, STAM.

ANS: I would say probability is the least important of the presented options. It is important for exams P and C, but I have not seen a ton of overlap in other tests.

ANS: Since Actuaries are in the industry of uncertainty, strong understanding of probability will help actuaries to price premiums, calculate and estimate claims, and reserve sufficient amount that will pay off any developing claims of insureds or policyholders. Probability is needed to be able to pass SOA/CAS Exam Probability, and STAM.

ANS: Probability is needed much more for actuaries. For example, loss ratio is the probability that a loss may occur.

ANS: Exam STAM requires.

ANS: Actuaries need a solid understanding of probability to calculate expected values, more specifically, Actuarial Present Values. These numbers are of special interest to regulators, who want to ensure insurance companies are solvent or to ensure that pension plans are well funded. Ultimately that is in the best interest of the public.

ANS: In the insurance industry, actuaries deal with the law of large numbers in order to mitigate risks. Mortality, morbidity, loss ratios, etc., are all concepts of probability and well taught in the Probability exam.

ANS: Actuarial work is all about calculating the chance of something happening (which is probability) and then the financial impact of it. The first exam is all about probability, which then lays a good foundation for the rest of the exams.

ANS: It is important to understand the various concepts around probability distributions. For example, the integration of survival function gives the limited expected value of such distribution. The limited expected value is an important concept in both life (life expectancy) and general insurance (expected size of loss) pricing.

ANS: Probability is the fundamental backbone of what an actuary does, understanding the range of Ultimate/Reserves. Providing on a singular claim basis, how often/likely something is going to happen. Without probability theories, none of that is possible.

ANS: Probability is the backbone of an actuary's job – our understanding of probability is what separates us from data scientists and accountants. We use probability to assess risks that otherwise are difficult to analyze. The examinations use probability in its most basic actuarial sense—on the job, computers do most of these calculations, but it takes an actuarial understanding to be able to interpret and analyze the results.

ANS: Actuaries need probability to determine expected value, which is used for pretty much every CAS exam.

ANS: Probability is very important to an actuary. You'll be asked to guide senior management in business decisions, and it helps to give them your best guess and what the probability is for unforeseen risk.

ANS: Insurance is about probability.

Q: **Why do actuaries need statistics? Please give examples and relate them to the SOA or CAS examinations.**

ANS: Statistics are important in being able to explain data to nonactuaries, or those with a less quantitative understanding. Being able to provide the right statistics to the right audience to best explain data is crucial.

ANS: Understanding statistics allows actuaries to extract meaningful information from large datasets and explain on a higher level or simpler terms what observations are inferred from the data, like trends.

ANS: Statistics are needed to analyze the uncertainty of risks. They are needed in Exam P, Exam MAS-I (CAS), Exam MAS-II (CAS).

ANS: Actuaries need to understand how to use statistics to accurately price insurance products and calculate reserves. Exam P, STAM.

ANS: Statistics are another aspect that I believe are not used as much in the later exams. I see it used in Predictive Analytics a lot, but other than that, I have not used stats much.

ANS: Actuaries need statistics for determining the reasonability and materiality of data. It is the responsibility of actuaries to clean datasets using statistical modeling and machine learning skills to determine if any given dataset is reasonable to historical data of the products such claim data, premium data, and accounting or financial data. After the analysis, it is the duty of the actuary to make decisions about the data, for example, the difference between two data sources, need to reconcile to see the true difference and make sense to dataset under study. Some of the statistical regression concepts needed are linear, multiple, nonlinear, Bayesian, density estimation, and scatter plot smoothing, etc.

ANS: Statistics is needed more by modeling team because they need to use data to build models to give a score or ranking of insureds based on the performance.

ANS: Exams SRM and PA require.

ANS: Actuaries use regression to model behavior of the real world, fitting models to past data and creating predictive models.

ANS: When dealing with big data/research, statistical techniques are often-times used to develop models and determine population characteristics. Exam C equipped me with such skills.

ANS: Statistics is a big part of Exam MAS-I and MAS-II.

ANS: Solid understanding of statistical concepts is increasingly important in conducting general insurance pricing. It is important for practicing actuaries to distinguish between correlation vs causation, and also appreciate the variance vs bias trade-off.

ANS: Statistics are typically the outputs we provide management and clients for the project at hand. Without the help of statistics, it is impossible to present our findings.

ANS: Statistics are closely related to probability. Basic statistics, such as mean, median, and mode, are important concepts that are applicable in almost any context. An actuary may want to determine the average premium, or find out what the median claim size is, or find the most common premium mode—statistics enables the actuary to perform all these tasks with ease.

ANS: Actuaries need statistics to understand loss distributions and the concepts of the mean and variance. Means and variances of distributions are important in Exams P, MFE, C, and S, amongst others.

ANS: Actuaries need statistics to better understand and justify factors used in pricing. I used statistics to calculate a factor by running over 100 random simulations; calculated the mean, median, and standard deviation; and weighted them to develop the factor.

Q: Why do actuaries need the theory of interest? Please give examples and relate them to the SOA or CAS examinations.

ANS: Time value of money plays an important role in setting reserves, particularly for long-tailed lines of business where a claim can be settled 10+ years from when reserves are being set. This has become especially important during uncertain inflation periods such as 2022.

ANS: Discounting and present value from Exam FM is used to determine obligations like DBO (defined benefit obligation), PBO (projected benefit obligation), or APBO (accumulated postretirement benefit obligation) in the real world.

ANS: They need the theory of interest when calculating the present value of premium paid in each term in future and the present value of estimated loss. The theory of interest is important in exam FM.

ANS: Exams FM and IFM. Actuaries work with long term horizons and thus need to understand the economics behind how interest and financial instruments work.

ANS: Theory of interest is probably the biggest thing for my current track (Health Fellow). A significant part of my job is calculating reserves, how these reserves grow over time. To remain competitive, our company has to price aggressively knowing the interest on reserves will be another cash flow. Interest arises in FM, MFE, LTAM but especially so in designing and pricing and finance and valuation of the FSA.

ANS: In order for any actuary to understand the insurance business as a whole, he or she will need to understand the theory of interest such as effective interest rate, effective rate of discount, nominal rate, nominal rate of interest convertible semiannually, monthly, quarterly, and continuously, variable rate of interest and many more. A strong understanding of interest theory will enable actuaries to evaluate present values and accumulated values of annuities, bonds, loans, interest rate swaps, and securities. Also, the theory of interest is the foundation or required skills for passing SOA Financial Mathematics.

ANS: I don't have the experience that requires the theory of interest.

ANS: Exams FM and IFM require.

ANS: Present value of annuities—calculate reserves for insurance reporting.

ANS: The theory of interest is frequently used when converting multiple-year projections to the present value basis, calculating annuity payments, determining reserves/cash values, etc. I got familiar with the concepts in the Financial Mathematics exam.

ANS: Time value of money is huge, especially on the life side. $100 today is not the same as $100 from 10 years in the future and actuaries should know why. Exam FM is all about the theory of interest.

ANS: Theory of interest is most widely applied in life insurance pricing to quantify present value of cash flows involved in an insurance contract. A good understanding of theory of interest becomes important for general insurance actuaries as they assume more leadership roles (to appreciate the impact of long tail vs short tail lines) or work in international markets (reserves are discounted under International Financial Reporting Standards).

ANS: Calculating benefits in the future and values of different options, we need to be able to discount back to current value to make the proper comparison. This is what FM taught us with the various theories of discounting.

ANS: The theory of interest is used by actuaries almost daily in order to calculate present values, which are extremely important when producing financial results. Because certain insurance products are long-term risks, present values allow an actuary to analyze future cash flows in today's money. If an actuary needs to determine the premium for a life insurance product, they will need to take anywhere from 5 to 30 or even 80+ years of cash flows into account—discounting these cash flows at an interest rate allows the actuary to assess the risk on today's term, when a decision has to be made.

ANS: Actuaries need theory of interest to determine the value of money across various time periods. Interest is a necessary component of finding present value, which is a huge component of exam FM.

ANS: Actuaries working in the financial area will use this more and in retirement.

Q: Why do actuaries need mathematics of finance? Please give examples and relate them to the SOA or CAS examinations.

ANS: Actuaries are important in an insurer's financials. Understanding management's goals and targets is crucial for an actuary. The actuary must command an understanding of basic finance to know what areas of the business to look at in order to better assess management's goals.

ANS: Exam FM teaches important concepts like discounting that is applied when determining present value of obligations for pension plans.

ANS: Basic knowledge for actuaries. Covered on FM and IFM.

ANS: Actuaries are embedded in the finance world. Understanding financial math is imperative to making proper business decisions.

ANS: Mathematics of finance, I think, plays an important role similar to the interest portion. I think it is a heavily tested portion of the exams, but the assumptions in SOA exams play out rarely in reality.

ANS: Since actuary profession is part of the corporate business which is made up of accountants, financial managers, CEO, stakeholder, and shareholders. At the end of the financial year the chief actuarial has the responsibility to communicate effectively to the various parties of the corporate world about state of the insurance business, such as how much reserves we currently have to offset any future claims. To communicate the financial statement in the context of reserving it is important for the actuaries to be able to interpret applications of mathematics of finance. Also, actuaries collaborate with professional accountants and financial experts daily, so it is very important for actuaries to have strong understanding of mathematics of finance in order to write yearend actuarial report that is aligned with financial best practices and to communicate effectively with stakeholders and managers. The mathematics of finance provides knowledge on how corporations operate, introduction to financial statement analysis, balance sheet, income statement, statement of cash flows, and financial reporting practices. The mathematics of finance concept will help actuaries pass SOA/CAS exam IFM.

ANS: I don't have the experience that requires mathematics of finance.

ANS: Not sure what mathematics of finance is....

ANS: Actuaries can also engage with finance/investment departments on corporate-level business plans, appraisals, etc. Understanding essential financial concepts will help them speak the same language as the other departments and communicate more efficiently. Exam MFE serves such a purpose.

ANS: Actuaries need to know what the numbers they calculate mean for the insurance company. This way they are able to give management a fuller picture instead of just throwing numbers in their faces.

ANS: Concepts of option pricing are usually not applicable to general insurance other than a few edge cases such as crop insurance.

ANS: At the end of the day, the presentation of our findings must be related to some financial metrics, how often loss is going to happen (frequency), how much loss costs in a certain layer, what are the maximum losses we are able to handle. With this information, treasurers are able to make informed decisions as to what they need to purchase.

ANS: Actuaries often work closely with the finance departments at their companies. When pricing a new product, management expects to see financial projections, like an income statement, showing the benefits of launching said product. And a reserving actuary needs to understand where those reserves are going and how they're developing and affecting the company's bottom line. I feel the SOA and CAS preliminary exams do a poor job of teaching actuaries finance and accounting.

ANS: Mathematics of finance are needed by actuaries to understand loans and annuities, which are important for Exam FM and MLC, amongst others.

ANS: Actuaries need mathematics of finance when calculating what trends are. If you're calculating what medical trends should be used, you would use what you learned in mathematics of finance to calculate year-over-year medical cost using trends.

Q: **Why do actuaries need economics? Do you think the syllabi of VEE-Economics provide enough coverage of economics for your work? Please give examples.**

VEE in the question is the abbreviation for Validation by Education Experience. Actuarial students are required to acquire knowledge in economics, finance, and accounting by, in general, taking approved college-level courses.

ANS: We need to understand how economics impacts insurance indirectly. How do oil prices change auto losses? How does inflation impact homeowners' loss?

ANS: VEE-Economics—understanding macro- and microeconomic picture helps to advise clients. For example, if we are in a year where you have a recession, we can anticipate that clients might not be open to contribute more to their pension plans compared to other years.

ANS: Pricing and reserving need economics knowledge and actuaries need to know economics to have a good business sense which is very important in work. VEE-Economics is enough.

ANS: Economics can help explain the driving forces of the overall industry of which actuaries are one sole part. We are given credit for VEE-Economics for essentially free. More economics cannot hurt.

ANS: I think the economic portion of the VEE is sufficient. The cycle of insurance is very determinant on seasonality and employment. You may want to hold more reserves for certain months because of deductible seasonality. Also you may expect more disability claims during times of high unemployment.

ANS: Actuaries need economics in order to help the insurance business they serve and the financial managers to make informed decisions about the cost of risk and uncertainty that arises in the insurance industry. Yes, the VEE-Economics syllabi provide enough coverage of economics for my day-to-day duties.

ANS: Actuaries need basics of economics, for example, supply and demand balance. When supply is larger than demand, the insurance industry is in the soft market; otherwise it's in the hard market. I think the VEE is enough so far.

ANS: Yes, actuaries need basic knowledge of economics.

ANS: Macroeconomics affects the inputs that are used by actuarial models, (e.g., interest rates). Hence, it's important for actuaries to understand how the economic landscape could affect their model output.

ANS: Projections on assets backing up reserve and surplus are closely tied to economics. Actuaries need to have basic knowledge about how each type of asset works, asset portfolio construction, etc., to gain a comprehensive picture of the projections. The demand and supply curve also facilitates reaching the optimum pricing. I feel the syllabi of VEE-Economics provide enough coverage, but I need to go back to the course when applying specific concepts to work.

ANS: The broader economy has financial impacts to an insurance company which need to be taken into consideration by an actuary. The syllabi of VEE-Finance help with this.

ANS: Economics is helpful when an actuary advises underwriting to make pricing decisions in a competitive insurance market. For example, in a niche inelastic market, a simple push for higher rate change might be sufficient to improve profitability. But in a market with elastic demand, better risk selection rather than rate push might be more beneficial.

ANS: I guess the most important theory from economics in the world of actuarial is the cost and benefits analysis, at what level we purchase insurance to give the best outcomes for our customers (from a broker perspective), whether it is to protect the working layer, or the catastrophic event that might strain the company's financials.

ANS: A strong understanding of economics is important for anyone in a business setting, not just actuaries. I felt that because my university's economics classes were good classes, the Economics VEE worked well for me. However, I cannot say the same for other universities—leaving it up to the schools to teach economics means there will inherently be an inequality in outcomes, as compared to subjects tested by SOA/CAS exams. Some teachers are hard graders or teach more than is required—the exams are at least consistent in their grading and content.

ANS: I think actuaries need to understand basic level of economics if they want to understand interest rate effects on products.

ANS: Economics is needed to understand inflation and supply/demand. I think the syllabi provide enough coverage since I barely use supply/demand in my day-to-day work.

ANS: Actuaries need to know basic economics to understand strategy in place. When a company decides to sell new insurances it may be because there is a need for insurance in a specific area.

Q: **Why do actuaries need accounting? Do you think the syllabi of VEE-Accounting provide enough coverage of accounting for your work? Please give examples.**

ANS: Actuaries use balance sheet information to assess financial goals and targets.

ANS: VEE-Accounting is mostly about credit and debit which I barely use on the job. Instead understanding frameworks like US GAAP and/or IFRS and how they apply to actuarial items like IAS 19R or ASC 715 (both terms describe accounting guidelines for employee benefit plans—authors' notes) would have been more helpful.

ANS: Actuaries need to read financial statements in everyday work. Accounting is a basic. VEE-accounting is enough and CAS actuaries also need to know Statutory Accounting principles used in insurance companies. This is covered in exam 6.

ANS: Actuaries need to deal with constantly evolving accounting regulations which fundamentally change how insurance companies calculate reserves. VEE-Accounting does not prepare someone to read accounting regulations.

ANS: I think accounting is accounted for in the VEE; however, I do not think it's something you can fully grasp until you work. Actuaries need to know the amount of monthly reserves set to make sure they remain liquid enough.

ANS: Yes, actuaries need basic accounting concepts to understand the details of financial statement, and we help in creating the financial statement by providing the accounting officers with the creditability of the dataset the accounting officers provided to us. Also, actuaries provide the validation or reasonability of the accounting records before the financial statements are developed and distributed to the stakeholders. I believe that VEE-Accounting provides the basic knowledge just needed by the actuary to perform his/her daily duties.

ANS: I don't have the experience in accounting side, so I think everything in the VEE-Accounting is enough.

ANS: Actuaries need to understand financial statements.

ANS: My knowledge of accounting came in handy when I had to analyze insurance accounting statements that were being projected by the actuarial model.

ANS: Basic accounting knowledge is essential to understanding the financials/proforma of a company/product. For insurance specifically, actuaries need to be familiar with statutory/GAAP/tax accounting reporting requirements. I don't think the syllabi of VEE-Accounting provide enough coverage of accounting for my work, especially for the characteristics and differences among the three reporting regimes.

ANS: Actuarial teams receive financial information from the accounting department. It helps with communication if basic accounting concepts are understood by the actuary.

ANS: Actuaries will need to understand insurance accounting and financial reporting standards to perform duties in reserving or conduct benchmarking via competitor results. A general understanding of GAAP accounting is helpful when studying insurance-specific accounting and financial reporting standards.

ANS: Exam 6 is fully in tune with actuarial accounting, so even though the VEE-Accounting provides basic concepts, Exam 6 is where the bulk of the learning came from. Accounting from an actuarial perspective ensures our ratios are in line with IRIS (Insurance Regulatory Information System) and helps us understand some early signs of financial troubles that we need to watch out for.

ANS: Accounting is important for pricing and reserving actuaries. When pricing a new product or setting a reserve, an understanding of accounting means the actuary will be able to properly assess the impact of the product on the company's financials. Actuaries need accounting more than, say, data scientists because actuaries are much more directly involved with the financials of the company. When I was in college, I didn't believe the Accounting VEE existed—I thought it was just economics, math stats, and corporate finance.

ANS: Actuaries need accounting to understand debt, assets, and liabilities. The syllabus for accounting provides enough coverage, since the exam 6 material isn't too much more advanced.

ANS: Actuaries need accounting to understand full picture of what is going on with the company's financials. For example, it is important to understand where the reserves set by the actuary go once accounting books your numbers. You'll be able to better explain variances.

Q: **Why do actuaries need finance? Do you think the syllabi of VEE-Finance provide enough coverage of finance for your work? Please give examples.**

ANS: VEE-Finance gives you an introduction to how the corporate world works especially about the hierarchy. Understanding finance helps you as an actuary working in deals since multiple parties are involved and today there are so many financing options such as IPOs or SPACs.

ANS: Actuaries also need to know finance since it's related to financial statements. Actuaries need to know companies' financial assets. VEE-Finance is enough.

ANS: Finance underlies all business decisions. More finance cannot hurt.

ANS: The VEE in finance has been adequate for my job.

ANS: Actuaries need basic concepts of finance to understand the details of financial statements, and we help in creating the financial statement by providing the financial officers with the creditability of the dataset the accounting officers provided to us. Also, actuaries provide the validation or reasonability of the financial records before the financial statements are developed and distributed to the stakeholders. I believe that VEE-Finance provides the basic knowledge just needed by the actuary to perform his/her daily duties.

ANS: I think actuaries need to understand basic finance. The VEE-Finance is enough.

ANS: Actuaries tend to work closely with the finance department. In many cases, actuaries act as advisors to the CFO who's ultimately responsible for decision making. Actuaries need to understand how their numbers are being used so that they can properly calibrate their models.

ANS: Actuaries need finance to understand how liabilities and surplus are financed as well as corporate financials. Financial metrics are also useful when communicating profitability, measuring liability characteristics, etc. I think the syllabi of VEE-Finance provide enough coverage of finance for my work, but I was not able to really master/memorize the concepts until I applied them to real work.

ANS: Actuaries need to know how to read a company's financials which the syllabi of VEE-Finance help with.

ANS: Financial concepts are important when actuaries advise leadership regarding profitability of various businesses.

ANS: Since a lot of the claims paid are in the future, a lot of times we do duration matching or some type of matching to allow us to have enough money to pay for claims while generating investment income to enhance the balance sheet.

ANS: Finance is very closely related to accounting for an actuary. An actuary needs to be able to understand the financial impact of an opportunity or decision and help guide management towards the best outcome. My corporate finance VEE experience in college was not very good. The class was disorganized, and the university's curriculum for the actuarial science degree at the time pushed us right into corporate finance without any introductory finance background. I wish there was an SOA exam that I could've studied for instead.

ANS: Actuaries need finance to understand credit and credit risk. SyllabI does provide enough coverage, since it gives the basics which can be learned more about in the exams.

ANS: Actuaries need finance since as an actuary in an insurance company you'll be in contact with the finance team to discuss profit and loss.

Q : Why do actuaries need risk and risk management theory? Please give examples and relate them to the SOA or CAS examinations.

ANS: Actuaries work closely with company management. We need to understand management's risk appetite to be able to best help in achieving company goals.

ANS: Understanding risk helps us advise our clients. We are able to look at their pension plans from different angles and advise them on how to mitigate those risks. For example, if a company is about to terminate their pension plan, we would advise them to speak to their advisor and get a more conservative approach for their investments.

ANS: Risk management theory is a basic in pricing and reserving. It's covered in exam 6 and online course 1 in CAS track.

ANS: Actuaries need to manage risk. Understanding risk and risk management theory is imperative.

ANS: Risk and risk management are integral to my job, and I think the largest aspect of it. You have to balance the profitable years with the unprofitable years. It is also important to price off trend rather than price off current year. The FSA exams teach you about rush hush crush which is something you see play out often.

ANS: One of the many actuaries' responsibilities is to minimize the cost of risks that a particular product or policyholders will pose to the business. Actuaries uses the risk-exposure-based cause and other factors to price a particular product so a strong knowledge about risk and risk management theory would be very important to identify, manage, and minimize the risk. Strong knowledge of risk and risk management theory will help actuaries to pass SOA MAS-II and Exam 5.

ANS: Insurance companies deal with risk all the time. Actuaries need to know risk management to help underwriting teams reduce risks to an acceptable level. The most common way in risk management of insurance company is through reinsurance. Actuaries need to understand the several methods in reinsurance, for example, treaty or facultative.

ANS: Actuaries need to understand how the mitigation of risks can affect probabilities and severities of loss, which ultimately affects the actuarial present values being calculated.

ANS: Actuaries work to mitigate risks. We need the risk management theory and framework to implement any risk mitigation solutions. In small- or medium-sized insurance companies, the chief actuary often serves as the chief risk officer for the company and is responsible for enterprise risk management. The ERM module is helpful with basic concepts, but the SOA ERM exam is a bit too technical (which I did not end up taking).

ANS: Statistical pricing methods are most useful when applied to homogenous policies. Risk and risk management theory help actuaries identify situations to avoid or mitigate even though it might not be easily quantified.

ANS: Most actuaries work in the insurance industry, so understanding risk and risk management theory is a must. We have an entire module that simply talks about how various insurance concepts work (quota share, excess of loss, captives, etc.)

ANS: Risk management and actuarial science really are two sides of the same coin. The numbers the actuaries are crunching are driven by the problems identified by those who are looking critically at the exposures risk professionals identify.

ANS: Risk is an application of probability—will something happen, or will it not? Actuaries use risk management to determine whether a product is a good idea – if there is a high probability of fraud, that could have a negative impact on the company. If there is a high chance of claims, premiums will need to be very high, and may not sell well to low-risk policyholders, therefore driving up the loss ratio on the product. Risk is at the core of all these decisions, and an actuary needs a strong understanding of it in order to be successful.

ANS: Actuaries need risk and risk management theory to understand claims and the insurance cycle. Claims are a part of every CAS exam from Exam 5 onward, and the insurance cycle (hard/soft market) is seen in Exam 6.

ANS: Actuaries need risk management to understand basic risk management concepts. As actuaries we put a price on "unknowns."

Q: **Why do actuaries need loss modeling/actuarial mathematics? Please give examples and relate them to the SOA or CAS examinations.**

ANS: Loss modeling is important for many actuarial functions. For example, we can fit capped data for a given line of business to a curve and use this curve to predict losses at a higher limit if the insurer is looking to offer higher limits.

ANS: I do not use the material from former exam C at all.

ANS: They are needed in risk analysis and pricing. MAS I, MAS II, and Exam 5 cover this.

ANS: This is an actuarial-specific discipline that underlies the models used by actuaries on a day-to-day basis.

ANS: Loss modeling is very important for the FSA exams in the health track. The FSA exams are very focused on real-life problems, and how to deal with them. For example, pricing plans is always important to remain competitive. Actuarial work is more of an art than a science which is a difficult concept to understand at the student level. I always want black-and-white answers, where in reality you don't price new business the same as you would renewals.

ANS: The actuaries need loss modeling/actuarial mathematical concepts such as modeling, random variable, characteristics of actuarial models (parametric and scale distribution), continuous models, discrete distribution, aggregate loss model, frequency and severity with coverage modification, maximum likelihood estimations, and more. These concepts are key to pricing actuary in order to accurately price premium for any coverage in the insurance industry. These concepts will equally help the practicing actuaries to effectively perform their work well and aid in passing SOA exam STAM and CAS exams MAS I and II.

ANS: Something related to credibility may be needed on the pricing side since when we price a product, we need to make sure the data is credible before we use it.

ANS: Probably more applicable for Property and Casualty. I'm in Annuities and Life insurance, where the loss severity is known.

ANS: Actuaries deal with mitigating losses and loss modeling helps them project future loss patterns and better monitor experience against pricing assumptions. Life contingency concepts are particularly helpful to me as a life and annuity actuary. Most of the concepts are covered in Exam MLC.

ANS: CAS Exam 5 deals a lot with loss reserving/rate making.

ANS: Loss modeling is the foundation to most of the general insurance pricing concepts. Many of the methods discussed in the loss modeling textbook are still being applied in ISO (Insurance Services Office) pricing methods.

ANS: Loss modeling/actuarial mathematics relates to insurance of which we typically do not understand the full cost when we charge for it. Using actuarial techniques, we are able to make a projection to ensure premium collected is enough to pay for all the claims.

ANS: Loss modeling is our bread and butter. Like probability, it's what sets us apart from other business analysts. By specializing in actuarial mathematics, we put ourselves into niche expertise that combines the skill sets of many different fields.

ANS: Loss modeling is important in understanding aggregate distributions, and how those can be different than the underlying risk. Aggregate distributions are used in Exam P, C, and S, amongst others.

ANS: In healthcare actuarial modeling is constantly used. One example is when calculating what a company would need to reserve for other postretirement benefits such as medical benefits for retirees.

Q: What stochastic ideas and techniques do actuaries use? Please give examples and relate them to the SOA or CAS examinations.

ANS: We perform stochastic forecast in projecting future assets of pension plans.

ANS: Many scenarios for valuation can be generated stochastically. This gives actuaries the confidence that the level of reserves they set is good enough.

ANS: Stochastic concepts are important for real-life job work rather than tests. They are a way to provide confidence intervals (rather than best estimates). They are used very rarely in finance and valuation.

ANS: Actuaries use the stochastic concepts and techniques such as the Chain Ladder Method, the Expected Loss Ratio method, the Bornhuetter-Ferguson method, Frequency-Severity method, Random Access Algorithm (RAA), the Cape Cod method, the separation method, Overdispersed Poison model, and more to assess the uncertainty about an estimated reserve update or figure. Also, to assess the risk associated with estimated indicated ultimate amount. The stochastic concepts and techniques are needed to pass SOA exam STAM.

ANS: I don't have the experience that requires stochastics.

ANS: Regulations are slowly catching up to more modern methods of calculating reserves and recognizing that there is a wide range of possible outcomes in actuarial predictions. Principles-based reserving in insurance annuity products use stochastic interest rates.

ANS: In the insurance industry, the new Principle-Based Reserve ("PBR") requires valuation actuaries to use stochastic scenarios for certain interest-sensitive products. Regulations are trending towards stochastic modeling in order to track tail risks more accurately. Additionally, many types of research involve random walk, probability distribution, stochastic process, etc., to draw more credible conclusions. The concepts and techniques are tested in Exam C.

ANS: ERM/Capital Modeling has a lot to do with creating stochastic models. This is important in that it helps companies calculate the risk they take on. Higher-level CAS exams deal with this.

ANS: For certain general insurance pricing situations, Monte Carlo simulations are employed to understand the impact of aggregate limits in reducing the expected loss cost for large insured with a high expected annual claims frequency.

ANS: Identifying averages is typically not enough, so we develop frequency and severity distributions by running Monte Carlo simulations that provide ranges based on the distribution we come up with, not just point estimates.

ANS: Some actuaries use stochastic modeling to project various scenarios for cash flows. For example, take an annuity product. The annuitant gives a lump sum to an insurance company. The company then has to manage that asset and distribute payments for the remainder of the policyholder's life. In order to manage this asset, stochastic modeling can be performed to project the effect of various economic scenarios on the asset. The actuary could use anywhere from 50–100 or more stochastic scenarios to come up with a final result.

1.7   Complementary Disciplines

In addition to being good in mathematics, economics, and other scientific subjects, actuaries need to have a broad arsenal of other skills.

Q: **Which are the most important complementary disciplines for an actuary and why?**

ANS: Economics—being able to quickly respond to external economic factors and how those may impact reserves, rates, customer retention, etc. Being able to respond quickly and proactively is what actuaries should strive to do. Not to be delayed in our response.

ANS: Knowing programming/coding is super helpful especially if an actuary is dealing with large data sets.

ANS: Programming is an important complementary discipline as models get more and more advanced and automations are more and more desired.

ANS: I believe risk management and pricing are the most complementary. You have to balance the risk of pricing aggressively the first few years of new business, with the risk they are not good business to write.

ANS: The most important complementary disciplines for an actuary are mathematics, statistics, machine learning (data analytics tools like Microsoft Structured Query Language—SQL, Studio, VBA, and excellent knowledge of Microsoft Excel), finance, economics, and insurance. I believe these are the knowledge and skills needed to be able to perform the daily duties as actuaries.

ANS: I think all the courses are equally important to students because they provide the basic knowledge for an actuarial career. Some of them may not be needed due to the career path, but they definitely help on the exams or other career developments.

ANS: Critical thinking mindset.

ANS: Computer programming is an essential skill for the modern actuary. As the volume of data is scaling up, Excel becomes frustratingly inadequate at performing efficient calculations.

ANS: Accounting and finance for entry-level actuaries. Risk management, risk financing, corporate structures, etc., are essential when moving up in the career path to see the bigger picture.

ANS: Public speaking and communication are both very important skills for an actuary to have. Being good at these skills can really set you apart from other actuaries.

ANS: Statistics is the most important, as statistical modeling is increasingly applied in general insurance pricing and reserving.

ANS: Public speaking—no matter how good our models are, without the ability to properly communicate that to our stakeholders, nothing else matters.

ANS: I think risk management and anything that encourages discussion and critical thinking. One of the biggest issues I've noticed in my career is an inability to communicate effectively with people in different disciplines. Ultimately, you need to be able to solve these incredibly complicated problems, but more importantly, you need to be able to explain your results in a way that makes people who may have no mathematical background at all understand and agree to them.

ANS: Communication is of the utmost importance—an actuary can be brilliant and solve all the world's problems, but if they cannot communicate those findings effectively, it will be for naught. An actuary also needs to know how to use computers and technology—nowadays, just about every company uses massive data sets and special actuarial software in order to crunch their numbers. Knowing how to use technology is essential to a successful career.

ANS: The most important discipline I would say is Excel file management and version control. Having files laid out in a simple and easy-to-follow format is a very important skill to separate your work from your coworkers at the same level. Also, saving each update to a file as a separate version makes undoing errors or seeing progressions of files much easier to manage.

ANS: I would say be able to understand and explain your results. As an actuary you'll have to be able to explain complex assumptions and results to senior management and you'll have to justify your results. As an actuary there are different ways and answers you can get since your assumptions are key, so it is important that you understand what you did, justify your assumptions and explain in basic terms how you arrived at your answers. The modules help you with that.

ANS: IT, computer science.

Software skills

Here is what the survey respondents said about the importance of computer skills in general.

Q: **What software skills should actuaries have and why? Please give examples.**

ANS: Excel for sure. Coding in SQL, Python, RStudio are all very helpful.

ANS: Actuaries should know how to use Excel, PowerPoint, and other software like data processing or data visualization like Power BI or Power Query.

ANS: RES-Q (used in reserving), SAS (basic coding language used in many insurance company), Excel, Google Cloud Platform, DataRobot (statistical analysis).

ANS: Excel.

ANS: Excel is by far the most important. We work in SAS a lot, but a lot of assignments have to be presented in Excel so external teams can understand.

ANS: The software skills or the data analytic tools in which actuaries should have strong understanding are Microsoft Structured Query Language (SQL), RStudio, VBA, and excellent knowledge of Microsoft Excel. These skills and tools are very important in completing any actuarial tasks.

ANS: Excel. We use Excel every day. Knowing Excel is really a basic skill at work. The functions we use all the time are Vlookup, Index + Match, rank, and pivot table. In addition, having the knowledge of SQL is a must during work. We pull data, upload data, and manipulate data daily.

ANS: Excel, VBA, some knowledge about BI tools.

ANS: Python is coming out to be a popular language, is open sourced, and there are a ton of libraries out there that add useful functionality, e.g., *pandas*. It's fairly easy to pick up and is very efficient.

ANS: Microsoft Office, especially Excel (VBA, Power Query). Analytical tools and programming are definitely a plus when analyzing a large amount of data. Actuarial tools (pricing/valuation/reinsurance, etc.) are often picked up at work.

ANS: Excel, Word, PowerPoint, Outlook.

ANS: A general understanding of SQL is important. It is also helpful to understand how to program.

ANS: R, Excel, and any kind of statistical software that allows the manipulation of large amount of data is required in the current world since data is king.

ANS: Being able to build a model that does your work for you may be a larger upfront investment, but it truly pays dividends in a field as cycled as insurance. A lot of the time, you are solving the same problem so to be able to automate any portion of the calculation is massive and an easy way to make yourself stand out. Most of the work is done in Excel, but R has been pretty popular in the companies I've worked for as well.

ANS: Actuaries should have strong Excel skills, and they should be familiar with at least one programming language—VBA is an easy one to start with and is practical, but R and Python are much more powerful and useful for larger data sets. R can also be used in conjunction with Markdown to produce reports and dashboards that can be used for in-depth analysis. I commonly use R at work to pull claims data, and I use packages like *dplyr* and *ggplot2* to break the data down and look for trends. I also use claims data alongside premium data in R to calculate loss ratios and see where our business is performing well. Many actuaries will learn about actuarial modeling software on the job – for example, in life insurance, many programs, like MG-ALFA and GGY-AXIS, function in essentially the same way. Learning one makes it easier to pick up another one.

ANS: Excel skills are the most important to have because Excel is the main program used in most companies. In both of the companies that I've worked in, I would say 90% of my work is in Excel.

ANS: Actuaries should know how to use Excel and SAS. As an actuary you would use SAS to pull data and you use Excel to model and create formulas for pricing and reserving.

Programming skills

Here is what the survey respondents said about the importance of programming skills in particular:

Q: Which programming languages do actuaries need and why? Please give examples.

ANS: SQL, Excel VBA. SQL is all about data storage. Data is important to actuaries. Being able to get the correct data quickly will help an insurer stay ahead of the game. VBA for automation purposes. Let actuaries solve other problems while the computer does the annoying work.

ANS: Some of the programming languages would be Python or R.

ANS: SAS and R are used a lot. SAS is powerful in analyzing data, and R is a powerful statistical language.

ANS: R, Python, SQL, VBA—R is used for exams, Python is a standard data science tool, SQL and VBA are found in various actuarial tasks.

ANS: SAS and VBA are two programming languages that can make your life much easier. SAS has the ability to handle and manipulate large claims or eligibility data sets. VBA can automate many monotonous tasks.

ANS: The programming languages actuaries need include writing effective query statements to pull important and necessary queries from given database. In order to write a good query, actuaries need to know the functionality of Structured Query Language. Actuaries need to be able to write and understand codes from RStudio software, and VBA. To be able to write programming languages, actuaries need to know functions, loops, recursive, abstractions, and nested, as well as datatype.

ANS: R or Python. Sometimes we have models to maintain, and they were built in R or Python. Besides, R is very powerful in statistics. VBA. A lot of functions in Excel are customized in the VBA. Knowing VBA is really helpful to understand the codes and revise them if necessary.

ANS: VBA, R, or Python.

ANS: I am not required to program at work (except for VBA), but I do see the trend in younger actuaries for research and analytics. Programming can certainly make certain tasks less labor or time intensive. From what I heard, Python is flexible and can be helpful. Oftentimes it depends on whether other actuarial colleagues have a preferred programming language to make peer review easier.

ANS: SQL, Python, and/or R.

ANS: RStudio, SQL, VBA.

ANS: Either R or Python is a good language to learn. As most of the widely adopted modeling methods are available in R and Python. There have been a lot of collaborative work in the open-source statistical community that allow R or Python to leverage each other's strength. As a result, it would likely be sufficient to be proficient in one of them.

ANS: Python/R/SQL—again this is with working with large amounts of data.

ANS: Actuaries should know at least one programming language on a basic level. VBA is easy to learn and useful for automating tasks in Excel, such as reading data files and summarizing them in a pivot table. R and Python are extremely powerful and, while slightly more challenging to learn, offer way more possibilities and data handling capabilities. R can handle files that are multiple gigabytes in size—Excel would never be able to handle files that size. RStudio can even handle multiple programming languages in the same Markdown document—an actuary can use SQL to read data, then use R or Python to massage it and analyze it. An actuary who is familiar with Markdown and R can produce beautiful reports that are easily maintained and updatable, and don't run the risk of being accidentally changed like cells in an Excel file.

ANS: Actuaries should always know Excel, R, and VBA. Any other languages will be a plus, but in my job these are the most relevant.

ANS: Actuaries need to understand at least SQL. SQL is required since most companies use it to store all of their data, and if not, then companies use SAS or Access where SQL code can be used as well.

ANS: SAS and SQL are important in healthcare. Some companies have SAS enterprise guide that helps create queries without knowing code, but some people just code, so it is important to at least learn how to read code.

Q: Do actuaries need to know machine learning, artificial intelligence, or data science in general? Why or why not? Please give examples.

ANS: It depends if the actuary's work relates to predictive analytics.

ANS: Data science is needed because most of the time we need to analyze data and do statistical analysis and build models. Machine learning is useful in modeling. Artificial intelligence is not what I am familiar with but probably will be of more importance in future.

ANS: No not really, it is not used in day-to-day work unless they are highly specialized.

ANS: I have not used these in my job.

ANS: Actuaries with knowledge in machine learning, artificial intelligence, or data science in general have an upper advantage over other actuaries. This is because financial and insurance industries hire experts in abovementioned areas to manage and maintain the database of the insurance companies. So having strong machine learning, artificial intelligence, or data science knowledge provides actuaries with a very strong advantage in understanding complexities of all the areas of the insurance industry. An actuary with those skills will be able to understand the functionality of the company's database, maintainability of the system, and create the system or improve the system for effectiveness and efficiency.

ANS: In general, yes. The insurance industry is changing. There are many AI insurance companies. Actuaries should keep an open mind and learn about the new technologies.

ANS: Soon, but not quite yet. There are still a lot of regulatory challenges to overcome before AI can be used for insurance.

ANS: It depends on which career path the actuaries enter into. Such skills are certainly critical for those in research and those dealing with large amounts of data.

ANS: Knowledge about machine learning, artificial intelligence, and data science is definitely beneficial to the actuarial profession.

ANS: While actuarial science and data science have concepts in common, they are still two different disciplines at the end of the day. I don't believe it is necessary.

ANS: Actuaries should have knowledge of how these methods work. These are powerful tools that can do a lot of harm if not applied properly.

ANS: Traditional actuarial work with reserving might not require as much machine learning/artificial intelligence of data, old concepts still work well. But on a ratemaking basis, in order to beat the competition, any way to identify variables to separate good from bad risk is a must-do.

ANS: Yes, as an actuary you need to know and understand the tools that will eventually replace the more repetitive actuarial tasks. That is how you can ensure your own value, by staying ahead of technology and supporting it.

ANS: It's important for an actuary to understand data, but the line between "understanding data" and "data science" is very blurred. I see the modern actuary as going down one of two paths—a technical, data-heavy path, or an analytical, finance-heavy path. One is not better than the other, and which path an actuary should take really depends on where that actuary's interests lie. These skills can be helpful at the right company, but they are by no means necessary for all actuaries.

ANS: I think it is not a hard requirement but will help your career in the long run. The set of skills would help with experience studies and modeling-type needs. And if you can enter in with that info, it would help, but it can be learned over the lifetime on the career rather than as a prerequisite.

ANS: Actuaries, in my experience, don't need to know it, but it is preferred. I haven't had any experience with machine learning, but I'm sure it's used at some companies.

ANS: In healthcare I haven't experienced much of that.

Business skills

It is often said that good business skills are essential to succeed in the actuarial world. Many companies now specialize in the teaching of business skills. It is also said that you need luck to succeed in life. But what is luck when it comes to an actuarial career? Seneca, the Roman philosopher, has provided us with a definition of luck that is quoted by many business schools today: *Luck is when preparation meets opportunity.*

Preparation in the context of an actuarial career is fairly easy to define. Becoming an Associate and then a Fellow of one of the actuarial societies comes close. But that is not enough.

Most actuaries are good in mathematics and most have many of the qualities considered essential. But almost all successful actuaries have learned to survive under enormous pressure, the pressure to succeed in examinations while working full-time and the realization that you don't have to ace the examinations; all you need to do is pass them. Keep this in mind as you enjoy the challenges presented in this guide.

Many, if not most, of these skills can be learned. Business schools specialize in developing these and related skills, which are discussed on the websites of most business schools.

Q: **What business knowledge and skills do actuaries need and why? Please give examples.**

ANS: Soft skills, such as the ability to clearly communicate, time management, working with a team, etc., play just as important a role as the mathematical and statistical knowledge and skills early in an actuarial career.

ANS: Communicating complex ideas or concepts in simpler words.

ANS: Excellent business sense with knowledge of finance, accounting, and economics is needed. And a knowledge of the insurance industry as well as the knowledge of insurance regulations.

ANS: Understanding insurance products and regulations.

ANS: I think communication is important. It is simple to communicate results to other actuaries. It is a lot more difficult to put terms such as loss ratio and IBNR in terms that can be understood by finance people.

ANS: Actuaries need to know how the business world operates. For example, in the Berkshire Hathaway Specialty Insurance company I work for, I need to know the regions we operate in, issuing and credited offices in the regions, the types of products we offer, the legal papers we write business on, such as National Indemnity Company (NICO), National Fire and Marine (NFM), Berkshire Hathaway Specialty Insurance (BHSI), etc. I need to know the product lines such as first-party insurance like homeowners insurance, and third party coverage like auto liability insurance, their PLS (Product Line Subtype) codes, and so many more. In a nutshell, having actuarial skills and full knowledge of how the company operates makes the actuary's job easy.

ANS: As an entry-level actuary, you need to be curious in the industry and career. You need to be eager to learn, to ask questions, and also to be responsible and careful with your work.

ANS: Actuaries need to keep a line of sight to the wider company and to understand how the products they support are making money for the company.

ANS: Time management, multitasking, proactivity, etc., for entry-level actuaries to help them settle in the actuarial world. Managing up, strategic decision-making, leadership, networking, etc., will help when moving up along the career path.

ANS: Risk is an important topic for actuaries to know. Calculating numbers only gets you so far. Knowing what to do with those numbers is more important and risk management knowledge can help with that.

ANS: Actuaries need to have listening skills to acquire business knowledge as they interact with underwriting. Many times, actuaries might not be able to address concerns of the business side until they modify the pricing model to be more in tune with the business reality.

ANS: The constant need to learn about the business you are working with. No one business is completely alike than another, and more importantly no one company is similar to another, so the business knowledge required is to learn the company you are working for and provide the best outcome for that company, and don't expect what works for a similar company will work for yours due to potential cost concerns, risk appetite, and other consideration.

ANS: Communication is huge, but a problem I have seen arise is a lack of a "commercial" mindset. I think actuaries in general tend to be very conservative in their estimates, and though I believe that is the correct place to start, in order for an insurance company to survive and make money. Being a stickler for the loss curve and hardliner in your model assumptions is only going to make it difficult for you to work with people in the long run. Stay open-minded and learn when to push and when to be flexible, and it will help you a lot.

ANS: I consider actuaries to be businesspeople who like math—not mathematicians who like business. At the end of the day, the actuary needs to understand the business they're in, and what its goals are. For example, an actuary who works on a specific product needs to understand whom the product is being sold to, and how they want to purchase the product. A renter's insurance product marketed to elderly people probably will not do well if the product can only be purchased via an app. And a life insurance product marketed to millennials and Gen Z will not do well if the only way to purchase the product is over the phone or in person. Actuaries also need to know how to conduct themselves professionally and with respect for others. Actuaries get a bad enough rap for not being the most social group—if an actuary doesn't behave in a professional or ethical manner, there can be extreme consequences.

ANS: Actuaries need to know profit, specifically the combined and loss ratios. Most reports that I've worked on put these ratios in their presentations because it puts the output of projects into a meaningful value.

ANS: Actuaries need to know what strategies are set in place by the company. This knowledge helps actuaries when developing assumptions.

Communication skills

Here is what the survey respondents said about the importance of communication skills as their careers unfolded:

Q: What communication skills do actuaries need and why? Please give examples.

ANS: The ability to effectively communicate work and findings to the level of stakeholders. Some audiences of actuarial reports and studies will not have a background in statistics, calculus, economics, etc. It is important for actuaries to be able to paint a picture clearly for all stakeholders to understand.

ANS: Communicating complex ideas or concepts in simpler words and using visuals to showcase large dataset. Using simple examples with numbers sometimes helps and pictures or slides explain a lot.

ANS: Actuaries need to deliver the result of complex statistical analysis to people with zero knowledge of this, which requires storytelling and an easily digested way of communication.

ANS: Being able to communicate and distill business problems to various stakeholders. Being able to convince stakeholders to support their ideas.

ANS: Actuaries need strong communications skills such as writing, reading, and speaking skills. This is because actuaries are expected to communicate their findings and results effectively to their managers and stakeholders. For example, a Chief Reserving Actuary like my manager should be able to communicate how and why he arrived at a specific ultimate loss ratio and ALAE value, why using a particular reserving method to complete any reserve updates is better than other methods, and much more. For my manager and any actuaries to do that he will need very strong analytic and excellent communication skills.

ANS: You should be able to explain things well. Make some examples when you answer questions. Be friendly and nice when you talk to other people.

ANS: Explain complex actuarial concepts to someone who has little knowledge about insurance.

ANS: Actuaries need to be able to break down complicated topics for their audience, which could range from HR (for pensions and health) to finance stakeholders.

ANS: Concise and clear communications with various internal and external stakeholders: peers, managers, senior management, and clients. Customizing communication style with audiences of different backgrounds. Making good presentation slides is also a plus.

ANS: Actuaries need to know how to effectively get their point across. Management has limited time and attention spans which makes good communication skills very important.

ANS: Actuaries need to be able to explain mathematical and statistical concepts in a way that most people can appreciate and understand.

ANS: Ability to concisely describe abstract concepts in layman terms. A report of many technical terms and thousands of pages of backup for the work you did is great, but no one is going to read through that. The lack of ability to distill it to a five-minute presentation to convey the most important finding is not going to be helpful for presentation purposes.

ANS: Communication skills are one of the easiest ways to stand out in your field. Being able to present to large groups and also collaborate with smaller groups will set you apart.

ANS: Actuaries need strong written and oral communication skills. In their day-to-day work, they will be on the phone or in-person with their coworkers often, discussing problems and responsibilities, or even just having casual conversations. They will also often write reports and emails that could go to the broader company, so they need to be able to effectively communicate their message to a wide range of people. For example, if an actuary is sending an email about the year-end financial results, they need to be able to cut down the information to only what's relevant and important, and they will need to get this message across clearly and effectively.

ANS: Actuaries need to know how to write clear, concise emails. This is because most people that you are emailing need to understand what the email is about in the first sentence. Also, actuaries need to be able to present projects in front of other people, so public speaking is important.

ANS: Actuaries need to know how to communicate through PowerPoint mainly when presenting to senior management.

1.8   Actuaries of the Future

One of the exciting aspects of an actuarial career is the continuously changing nature of the profession, and of the accreditation systems. As the world evolves, so do the role and required skills of actuaries in it. What will be some of the future skills required to adapt to this change? How do actuaries view the need for new skills as the profession evolves? In today's data-driven world, Schwartz and Douglas (2019) point out that people's day-to-day activities are coded into zeros and ones for use by organizations in an ever-expanding number of ways, and that today's unprecedented access to data is reshaping the capabilities of insurance actuaries. More reliable statistical data, including mortality data, are being actively collected. Over time, this new information will find its way into the actuarial world. New skills will be required to create credible predictive models based on these data. Knowledge management and global communications will continue to advance. International business will thrive and the need for foreign language skills will increase as a result. The globalization of manufacturing will create a need for new forms of insurance. The fusion of banking, insurance, and wealth management will create new actuarial challenges and opportunities. The list goes on. Equipped with the appropriate skills to tackle these challenges, actuaries will play a key role in assessing the risks inherent in these changes. Here is how some of the respondents to the survey see the impact of these and other changes on the future of the actuarial profession:

Q: **What changes in the knowledge, skills, and mathematical techniques in actuarial practice do you envisage in the next 5/10/20 years?**

ANS: Actuaries will be asked to have more coding skills out of school over the next decade. Companies are investing heavily in automation and streamlining their processes. Being able to collect, store, and leverage data continues to be the key. Coding skills will play a huge role moving forward. As things become automated, actuaries will need to find new ways to solve new problems. While one process becomes automated, actuaries will begin a new process. As such, actuaries will need more general business knowledge and understanding of how broader economic conditions will impact the business. Actuaries should break into other industries (beyond) over the next decade, so more knowledge on general business and broader economic conditions will be important.

ANS: Using data visualization like the metaverse to explain concepts instead of Power Points or Power BI visuals.

ANS: I think more and more data science, machine learning, and artificial intelligence will be used. Actuaries will research more on risks that are uninsurable for now and new risks that will occur in future.

ANS: Programming will become more and more prevalent. I don't expect models to get too advanced as they become a black box which regulators and auditors do not like.

ANS: I think the data will be significantly more automated. An unfortunate part of our job is reconciling data, but I feel there is an initiative to clean the data prior to receiving it.

ANS: The changes in the knowledge, skills, and mathematical, statistical techniques in actuarial practice I envisage in the next 5/10/20 years include how the transition from just conceptual understanding of those mentioned concepts and techniques to more advanced machine learning and artificial intelligence that will reflect the evolving changes in the world of insurance.

ANS: Knowledge is always the most important in the career and actuaries should grow it no matter what. All the skills or techniques may be needed less when you go to the manager level, but knowledge is something you can rely on forever.

ANS: Machine learning, maybe.

ANS: Computer programming is going to become a bigger part of the profession.

ANS: I would imagine an emphasis on big data/advanced analytical skills/ machine learning as these are also trends in the regulatory framework. In my Global Destination Course, I noticed these techniques were already mature in the European insurance industry, especially for audits.

ANS: Actuarial work needs to evolve in order for companies to be more competitive.

ANS: I think eventually Bayesian statistical approach would become more mainstream.

ANS: I think eventually the old way of using historical data to generate LDFs/ILFs will go out the door, and a more intelligent way will be available to use claims history to better forecast what the reserve needs to be.

ANS: I think a lot of the more standard actuarial roles (like underwriting and broking) will become more automated over time, meaning that more collaborative, problem solving roles for nontraditional problems will be the future.

ANS: I foresee many more calculations and processes becoming automated. It will be important for actuaries to understand what's going on behind the scenes of any automation; otherwise they will not be able to paint an accurate picture of the results. I don't foresee much change in the background knowledge provided by the exams—actuaries will always need to know probability and statistics and finance. What will change, however, is how much they use the technical parts of those skill sets versus how much they use the analytical parts. Where I do foresee some change is in the tools used for analysis—as data gets bigger and captures more information, some actuaries will try to blur the line between data scientists and actuaries, and they will use sophisticated predictive models to drive analysis and decision making.

ANS: More need for knowing multiple programming languages will be seen in the future.

1.9   SOA vs CAS

Traditionally SOA members specialize in actuarial matters of life insurance and health insurance, while CAS members work in property and casualty insurance.

Q: What is the main difference between working in SOA and CAS?

ANS: The CAS deals with property and casualty lines. There seem to be more lines of business covered by the CAS compared to the SOA, and these lines covered by the CAS aren't necessarily closely related. Workers' Compensation is much different than homeowners', which is much different from Offshore Energy.

ANS: I think CAS is more geared to P&C while SOA is more for the other actuarial fields.

ANS: I'm not familiar with SOA track.

ANS: The type of insurance products and models.

ANS: I do not have experience in CAS.

ANS: SOA designation helps actuaries acquire knowledge and skills for life insurance, health, pensions, and retirement, whereas CAS designation prepares actuaries for property and casualty insurance.

ANS: Most lines in CAS are short-term while SOA has a lot of long-term products.

ANS: I've only seen the SOA side.

ANS: I've only had experience in SOA (life and annuity), but in general, I think product cycles in SOA are longer than those in CAS, and product development/valuation methodologies are very different. Some say that working in CAS (P&C) is more exciting and innovative, but that's also true for SOA because the life product types and regulations are evolving rapidly, and we need to prepare ourselves for the new challenges.

ANS: SOA deals with life/health while CAS deals with P&C. Two different organizations and they both have their own benefits.

ANS: P&C insurance companies tend to insure quite more heterogeneous risks, and the possible actuarial work that can be done is more dependent on other parties, such as underwriting and claims.

ANS: Work is much more technical on the CAS side. You will have to calculate and to program a lot more. On the SOA side, you work more with people outside of the actuarial profession.

ANS: I had only worked on the Casualty side.

ANS: SOA is a bit less technical and more regulations-driven than CAS.

ANS: I think just the products. SOA is life and health, whereas CAS is property and casualty. Actuaries tend to focus on one of these two sides, and even within each side an actuary can focus on one or a few specific products.

ANS: CAS is based more on property and casualty risks/insurance, whereas SOA is based more on life/health insurance. Also, there are more career opportunities in the SOA than the CAS.

ANS: SOA for life and health, CAS for Property and Casualty.

Actuarial students also often ask, "What actuarial path, SOA or CAS, should I choose?" The typical answer would be, "Your employer determines." The truth is that in most cases the nature of the insurance business of a student's first employer determines the actuarial rookie's career path. Actuaries of life and health insurance companies are mostly members of SOA while members of CAS typically work for property and liability insurance companies. Consulting firms employ members of both organizations as their clients can be from both industries.

1.10 Accreditations

Q: **How important is the number of actuarial examinations passed for an actuarial career? Illustrate your answer with examples of your career and others in companies you have worked for.**

ANS: Achieving ACAS is important I would say. FCAS is not as important. The Deputy Actuary at my previous company was only ACAS and was recently promoted to Chief Risk Officer. I notice that individuals later in their careers who have not achieved ACAS are mostly individual contributors while ACAS/FCAS are team leaders and decision-makers. However, I know many who have stopped at ACAS, are excellent communicators, and have continued their way up the ranks of a company.

ANS: Having your credential helps you climb the ladder faster since you will be required at some point to start signing reports.

ANS: It is very important especially for higher-level exams like Exam 5 and up. Exam 5 has a lot of pricing and reserving practical knowledge. After I passed the exam, I felt it was easy to understand the process of pricing and reserving.

ANS: Extremely important, exams dictate your pay and role. Without credentials, you may be denied a promotion.

ANS: I think the exams are very important for jobs. In order to get promotions and move up, you have to show the ability to pass the exams.

ANS: The higher the number of actuarial examinations passed by an actuary, the higher knowledge and skills she has acquired and the better she can effectively do her job. Also, passing more exams provides the actuary with more compensations (higher salary) and bonuses by their employer.

ANS: My manager once told me that study was part of the work. The more exams you pass, the easier you get promoted.

ANS: You got salary increase and promotion with more exams passed and credentials attained.

ANS: In the pensions industry, an FSA designation is desired but not compulsory. One can be quite successful with an ASA as long as they also get their EA. In the life insurance company, an FSA is almost a must to move up within the company.

ANS: In my company, we typically would like to see intern and entry-level candidates with at least 2–3 exams to demonstrate their actuarial capability. After that, they should try to pass at least 1–2 exams a year to stay in the study mode. The number of actuarial examinations matters because not only are they tied closely to salaries and promotions, but because each additional exam equips actuaries with exactly what they lack at work.

ANS: In order to keep progressing in an actuarial career, passing exams is very important. Being credentialed helps with actuarial credibility. Leadership roles are given to those with ACAS or FCAS.

ANS: It depends on the type of company. For example, some big companies might reserve certain internal grade level and above for Fellows only, and an associate might need to be truly exceptional to cross the barrier.

ANS: I had been pretty lucky even though my lack of success in the exams (two shy of FCAS now) had not hindered my job prospects, but that's probably because I am not working for an insurance company.

ANS: Generally exams are more important for insurance carriers than consultants. Carriers will encourage actuarial students to finish their exams as quickly as possible and are very flexible with study hours. Consultants are less flexible and therefore less concerned with passing exams—they'd rather get their billable hours than study hours.

ANS: The number of actuarial exams passed is important in my opinion until the first level of credentials have been completed. That being said, the quality of someone's work as an actuary is more important than the number of exams passed. For example, the first company I worked for didn't particularly care if you passed exams or not, as long as you got your work done and you got better at your job.

ANS: From what I have seen, it is different for consulting and insurance. In insurance if you are just working in actuarial division, you need an ASA to be promoted to a manager or director and you need a Fellow to be promoted to VP or chief from my experience. In consulting, you need an ASA to be manager and above.

1.11 Associates and Fellows

Q: What can a Fellow do in your company that an Associate is not able to do?

ANS: Sign off on reserves if the associate has not passed exam 7 of CAS.

ANS: EAs can sign pension funding reports.

ANS: More strategic work.

ANS: Previously, Fellows could be managers and Associates could not.

ANS: The fellow takes on more independent work than an associate. The Fellows also deal directly with the chief actuary.

ANS: A Fellow can hold managerial positions in Berkshire Hathaway Specialty Insurance (BHSI) but an Associate cannot be a manager at BHSI.

ANS: Officially approve and sign actuarial memorandum.

ANS: An Associate would not be able to become the Appointed Actuary at the life insurance company I work for.

ANS: Sign certain certificates, attend certain industry events, and have "actuary" in the job title.

ANS: Fellows usually have more leadership positions, and I am pretty sure only an FCAS can sign the statement of actuarial opinion.

ANS: I don't think there is a strong correlation between an actuary's technical capability and success in passing exams. I believe most ACAS are just as competent as FCAS at the same grade levels in performing their actuarial duties.

ANS: Fellows command a certain level of expertise that Associates do not. Associates can be very experienced and know a lot about the business, but it's assumed that simply because a Fellow has taken more exams, they will have more knowledge. There is no real difference in responsibilities, though. Both can sign statements of actuarial opinion, manage teams, and drive business decisions.

ANS: I am not aware of any differences, seeing that an Associate can still sign off on reserve opinions.

ANS: A Fellow of SOA can be a chief actuary, but an ASA can't.

1.12 Going for a Master's Degree

In some countries like Australia, Canada, China, and the United States, many actuaries begin their pursuit of actuary in college majoring in actuarial science. Actuarial students take actuarial and related courses, begin to sit professional exams while in college, and take more exams on the job. In these countries, actuaries have to spend many years studying for their professional examinations. As a result, there is little incentive for acquiring a Master's degree. Master's degree in actuarial science programs is most beneficial to people who hope to switch their career to actuaries.

In countries where the education of actuaries is university-based, it is often customary to pursue graduate studies at the Master's level. Having a Master's degree is in some sense equivalent to having achieved Fellowship status in a professional society. A Master's degree is required, for example, to become an Appointed Actuary in a country like Denmark.

Here is what some of the respondents to the survey had to say:

Q : Discuss the value of graduate studies in actuarial science, accountancy, finance, economics, MBA, and so on, for an actuarial career.

ANS: I don't believe graduate studies are beneficial for an actuarial career. What you learn in your first few years working is more valuable than what you'll learn from a graduate program.

ANS: Most of the learning is on the job and through the exams. I think going for graduate studies does help unless you are getting prepared for the exams.

ANS: Graduate studies in these areas build up a solid foundation for a student to become an actuary because actuarial work is a combination of statistics, finance, economics, business sense, accounting, and data analysis. Graduate studies will give students opportunities to grasp knowledge in these areas.

ANS: They have marginal to little benefit. Exams and becoming credentialed trump graduate studies.

ANS: I think the graduate study in actuarial science helped me understand what I really want to do it. The actuarial exam track is a grind. It is easy to say that as an undergraduate student, but I feel I really understood it afterward.

ANS: Graduate studies in actuarial science, accountancy, finance, economics, MBA, and so on would help actuaries to have a full, complete, and holistic approach to the complexities of the insurance industry.

ANS: Not too much unless you are changing your career to the actuarial field. My managers are undergraduates, and they are happy with it. Knowledge and experience are the most important in the actuarial career path. Besides, there are so many exams in this career, there is no need to pursue another diploma if you are already in the actuarial career.

ANS: Mostly not needed. Maybe an MBA would be helpful as one aims to move into nontraditional roles, like in finance or moving up to Chief Actuary.

ANS: I believe an MBA degree is necessary for any senior management position, but not just for actuarial. For a pure actuarial career, I do not think the other graduate studies are as helpful.

ANS: I can see this being valuable for an actuary looking to move up in the company and taking on a leadership role. Higher education can help actuaries learn more about how to run a company as well as how actuarial fits in the big picture of the company.

ANS: I am not sure. It might help with acquiring leadership roles.

ANS: Actuarial Science provided me with a great deal of concepts in the math side of what I need to know in the world of insurance. My MBA in financial risk management helps bring those concepts together from a business sense.

ANS: I believe only two groups of actuaries would see value in graduate studies—nonactuaries who are considering changing careers to become actuaries and actuaries who want to work in higher education rather than industry. Otherwise, I believe an actuary's time would be better spent studying for actuarial exams, learning new skills on their own outside of work and school, or enjoying their time away from work.

ANS: I don't believe there's any value in graduate studies for a standard actuarial career, but if someone wanted to move to Chief Actuary or a higher management role then I'd imagine an MBA would be essential.

ANS: My actuarial background has enabled me to leverage my MBA degree into a very valuable career as a Management Consultant. My post-MBA career, however, is not related to the actuarial discipline.

1.13 Career Alternatives

What if you have spent many hours studying to become an actuary and at some point you simply say to yourself that it is time for a change? Is there anything else you can do with all of this specialized training and knowledge? Here are some alternative career options:

Q: **What are the alternative professional options for actuaries who decide not to pursue an actuarial career? Please give examples.**

ANS: Underwriting, risk management, or data analytics for other industries (airlines, banking, finance, etc.)

ANS: Teaching.

ANS: Data scientist. Most of actuarial skills are transferrable to data scientists.

ANS: Underwriting, data analyst, data scientist.

ANS: Data scientists and mergers. These let you apply your benefit knowledge and data knowledge to other aspects.

ANS: The alternative professional options for actuaries who decide not to pursue an actuarial career include enterprise risk management, data analysis, IT, MBA, accounting, finance, and teaching.

ANS: I know some actuaries who cannot pass exams have switched to data department in order to maintain databases for actuaries. Some people go into underwriting.

ANS: Underwriters, financial/investment companies, data scientists, etc.

ANS: Data science.

ANS: Risk managers, finance, underwriters.

ANS: Underwriting is always a good career alternative for actuarial majors.

ANS: Risk managers/underwriters/risk analysts/catastrophe modeling analysts.

ANS: I found my sweet spot in alternative risk transfer. Whether you are on the brokerage side or the underwriting side, you have to have a critical, analytical mind. Frequently, I would have to come up with new ways to quantify things that there was no rule book for, meaning I would collaborate very closely with our actuaries and other specialists to solve the puzzle. Every day is different, and that is what really drew me to the field.

ANS: Anything in risk is a great opportunity—either elsewhere in an insurance company or outside of insurance in risk management at a company. Many companies have risk managers that quantify and analyze their risks, especially on an enterprise level. Actuaries can also take their quantitative skills into just about any other finance-related career: accounting, investments, personal finance, etc.

ANS: Alternative options would be working for an insurance company as an underwriter, claims person, or broker. These all would work with actuaries in some sense, and all the concepts of risk still apply to these jobs.

ANS: Medical economics and underwriting. There are actuaries that work in these fields using what they know from being an actuary.

ANS: Investment banker, portfolio manager, accountant, statistician, professor, researcher, economist.

ANS: My friend stopped being an actuary and went to Aon. She works in benefits for Aon for mergers and helps crosswalk the plans.

ANS: Yes, one of BHSI team members switched from actuarial to become an IT technician in BHSI and then he left BHSI to join a start-up IT company in NYC.

1.14   Actuaries Around the World

As is the case with most formally structured professions such as medicine, engineering, accounting, and others, the international employment of actuaries involves two critical elements: a recognition of academic qualifications attained in another country and a license to practice. In Europe and Latin America, actuaries have tended to qualify by completing a course of actuarial study, usually up to the Master's level, at universities accredited by actuarial societies or governments, and by meeting certain professional requirements.

In Great Britain and Commonwealth countries, the Institute and Faculty of Actuaries has defined an actuarial syllabus and sets of examinations based on this syllabus. Students in many parts of the world take these examinations to become actuaries in their countries. Certain university courses at designated universities can be credited towards this process.

In the United States, the Society of Actuaries and the Casualty Actuaries Society have defined a syllabus and sets of examinations that must be taken to become an actuary. Most American and Canadian actuaries and many others around the world become actuaries by passing these examinations.

Regardless of the different accreditation processes, the actuarial skill set is universal. With the process of globalization of the world economy, the two systems move toward each other, as pointed out by Brown's paper on the globalization of actuarial education (see Brown (2002)). Several countries with a university-based accreditation system institute nationally administered examination systems, whereas countries with professionally based accreditation systems begin granting exemptions for some courses taken at university. The University Accreditation Program of Canadian Institute of Actuaries allows students of accredited programs to receive credits for their courses. Starting in Fall of 2022, selected university programs and courses are credited toward certain SOA examinations.

It is actually fairly easy for actuaries educated in North America and the United Kingdom to work in other countries. The following examples will give you some idea of different national accreditation systems and of the portability of the acquired qualifications between certain countries. All countries in the list that follows, in no sense a complete one, have actuarial association(s) as full members of the International Actuarial Association (IAA), and European countries have organizations as full members of the Actuarial Association of Europe (AAE).

Argentina

The education of actuaries in Argentina is university-based. It is necessary to obtain a degree of Actuario at a local recognized university in Argentina, like the University of Buenos Aires, the 21st Century University, and Universidad del Salvador, that follow the syllabus recommended by the IAA (details in Chapter 2).

The University of Buenos Aires issues diplomas in the following two orientations, namely Actuarial Administration and Actuarial Economics, each of which provides an actuary with a license to practice. The degrees differ only in some courses on administration and economics.

Each state in Argentina has a public professional council called the Professional Council of Economic Science (CPCE, Consejo Profesional de Ciencias Económicas), created by law, which controls the independent activity of accountants, actuaries, administrators, and economists. These councils are responsible for maintaining professional standards. In order to be able to issue independent reports and formal advice, actuaries must usually be registered with the Council of the state in which they work.

Registration requires an actuarial diploma issued by a recognized Argentine university (private or public) or a diploma from a foreign university recognized by a public university with a full actuarial program, such as the University of Buenos Aires.

The Professional Council of Economic Science of the City of Buenos Aires is the only actuarial association in Argentina, and it is a member of the IAA.

Australia

The Actuaries Institute Australia (AIA) is the professional body representing the actuarial profession in Australia. The actuarial education program in Australia is made up of three parts, the Foundation Program, the Actuary Program, and the Fellowship Program. The Foundation Program consists of six core principal subjects: actuarial mathematics, actuarial statistics, financial engineering and loss reserving, risk modeling and survival analysis, business economics, and business finance. All six subjects must be completed.

The Foundation Program can be completed by studying an undergraduate actuarial degree at one of eight universities—Bond University, Curtin University, Macquarie University, Sydney, Monash University, the University of Melbourne, the Australian National University (ANU) in Canberra, the University of New South Wales (UNSW) in Sydney, and Victoria University of Wellington, New Zealand. Alternatively, these subjects can be studied by correspondence through the Institute of Actuaries (London). Credit transfers from another actuarial society are also recognized. Students may be granted exemption from subjects based on university grades that are consistent with Institute standards.

The Actuary Program teaches candidates how to apply technical actuarial skills to a range of problems across different business environments. Two subjects taught by eight universities in Australia and New Zealand (as listed) are actuarial control cycle and data analytics principles. Two other courses delivered by the Actuaries Institute are asset liability management, and communication, modeling, and professionalism. A strong and rigorous policy framework for accreditation of the university courses is in place so that the Actuaries Institute maintains quality control of the teaching and assessment of the courses. After completing the Foundation Program and the Actuary Program, members achieve Associateship of the Actuaries Institute.

Those who have attained associateship must pass three Fellowship modules and complete a year-long supervised work experience to become Fellows. Each module allows candidates to choose one of four options ranging from pricing to evaluation, investment of life insurance, or general insurance products.

The AIA has concluded a number of mutual recognition agreements with actuarial organizations of other countries. Mutual Recognition Agreements (MRAs) allow the conferring of qualifications of the Actuaries Institute on applicants who are Fellows, or in some cases, Associates (or equivalent) of another Actuarial Association—in return for broadly similar conditions being applied by those Actuarial Associations to members of the Actuaries Institute in the corresponding membership classes.

The basic reasons why the Institute enters an MRA may be summarized as follows:

1. to facilitate global trade in actuarial services, and the mobility of individual actuaries, by providing criteria for the recognition of appropriately qualified actuaries from other organizations; and
2. to recognize similar qualifications, to avoid unnecessary barriers, and to enhance the global provision of education, research, and professional services.

The AIA has or has had MRAs in place with other Actuarial Associations including but not limited to:

Actuarial Society of South Africa (ASSA)
Canadian Institute of Actuaries (CIA)
Casualty Actuarial Society (CAS)
Institute of Actuaries of India (IAI)
Institute of Actuaries of Japan (IAJ)
Institute and Faculty of Actuaries (IFoA)
New Zealand Society of Actuaries (NZSA)
Society of Actuaries (SOA)
Society of Actuaries in Ireland (SAI)

Austria

The Austrian equivalent of a Fellow of the Society of Actuaries is a full member of the Actuarial Association of Austria (AAA). The main function of the AAA is to promote the education and training of its members, to represent actuaries both nationally and internationally, and to establish guidelines and rules for good actuarial practice in Austria.

To become a member of AAA, a candidate must have obtained a university degree in mathematics or an equivalent program, completed training in insurance economics and insurance law, and have 3 years of professional actuarial experience. The Technical University of Vienna is the only Austrian university offering novel degree programs in actuarial science following the North American Bachelor's and Master's degree structure. The Technical University of Vienna also offers courses on insurance economics, accounting, and related law.

Austria is a full member of the Actuarial Association of Europe (AAE), an organization representing actuarial associations in Europe that provides advice and opinions on actuarial issues in European legislation. The Mutual Recognition Agreement (MRA) coordinated by the AAE stipulates that qualified associations recognize the credentials and training of other qualified associations. AAA is a participating member of the agreement. See Appendix C for the full list of the agreement.

Belgium

The Belgian equivalent of a Fellow of the Society of Actuaries is that of a full member of the Institute of Actuaries in Belgium (IAiBE), a full member of the AAE and a qualified member of the MRA. The requirements for becoming a member of IAiBE include three aspects:

- Theory—The applicant must provide evidence of a thorough theoretical knowledge of actuarial sciences, including basic actuarial training and advanced skills as described in the IAiBE syllabus. Students who have completed studies in Actuarial Sciences at a Belgian University automatically fulfill this requirement.

- Practice—The applicant must engage in, plan to engage in, or have engaged in professional actuarial activities.

- Ethics—The applicant must undertake in writing to respect the Institute's Code of Professional Conduct and to submit to the Institute's Sanctions Policy.

Brazil

The education of actuaries in Brazil is university based and is offered only at the undergraduate level. The focal points are Rio de Janeiro and Sao Paulo. The profession is loosely organized through the Brazilian Institute of Actuary. The Institute represented Brazil at the first international professional meeting of leaders of the actuarial profession and actuarial education in Latin America, held in Buenos Aires, Argentina, in 2002. Brazil has no actuarial Fellowship system and certain aspects of life insurance and reinsurance are government run.

Canada

Canadian Institute of Actuaries is the actuarial organization that represents Canada in the International Actuarial Association. The two categories of membership of CIA are ACIA and FCIA.

In 2022, CIA announced the new ACIA Capstone exam and modules. Accredited university programs play an important role in the new qualification pathways. Accredited universities offer, as part of their degree programs, a set of mandatory courses that meet syllabus requirements established by the CIA. The CIA requires no minimum course grades and a program's education quality in training entry-level actuaries will be verified through the CIA-administered ACIA modules and ACIA Capstone exam.

AICA modules are intended to bridge accredited university education or education obtained via other actuarial organizations with the work environment. They are administered online via the CIA learning management system.

- ACIA Module 1
  Introduction and actuarial communications: The first portion introduces the profession and gives a general context to the candidate; it explains the history of the profession and describes the context of actuarial work and the different fields of practice. Communication in an actuarial setting is discussed: audiences, technical and nontechnical writing, and the impact of cultural diversity on communication.
  The actuarial environment: candidates will learn financial systems, the external forces actuaries are facing, and the risks stemming from changes in

these external forces. Risk management, quantification, mitigation, and transfer are explained in detail.

Actuarial work: candidates will learn how actuaries work to provide sound solutions. The actuarial control cycle is introduced, and each component of actuarial work is described. Decision-making and problem solving, and ethics and professionalism are introduced.

- ACIA Module 2

Traditional actuarial solutions: This portion of the module complements university or other recognized education by reintroducing actuarial solutions to traditional problems. Fields of application of such traditional solutions could be pensions, life insurance, or general insurance.

Advanced actuarial models and predictive analytics applications: This final portion of the module introduces predictive modeling to actuarial problems, building on the predictive analytics models taught at Canadian universities or through other recognized actuarial education to prepare candidates for the workplace.

There is no formal prerequisite for the ACIA modules. Candidates in an accredited university will be expected to have completed some actuarial courses prior to completing the first ACIA module, likely in their second year of university. The second module should be completed when a candidate has nearly completed all courses relevant to the ACIA Capstone Exam.

- ACIA Capstone Exam

The overarching goal of the ACIA Capstone Exam is to help candidates prepare for the actuarial profession and tasks related to entry-level actuarial positions. Candidates are expected to demonstrate integration of knowledge of actuarial concepts and communicate results with a few short assignments. The exam assumes knowledge acquired in a recognized university degree, as well as from the ACIA module, focusing on application and communication. The ACIA Capstone Exam is administered as an open-book, 6-hour exam, requiring analysis in the context of a problem and submission of written responses to specified tasks. The exam is split into two sections: a 4-hour mandatory common section and a 2-hour specialized section, long- or short-term, selected at exam registration. Candidates will complete the exam through the learning management system, with secure online proctoring.

A university-accredited degree and successful completion of the ACIA module is required background knowledge for the Capstone Exam. The background knowledge covered in an accredited university degree and expected of students ready for the Capstone Exam are outlined as follows.

A1. Probability
A2. Financial mathematics
A3. Business, economics, and finance
A4. Actuarial mathematics
A5. Predictive analytics

- FCIA

CIA offers five specialty tracks for fellowship options. Each track is comprised of two online FCIA modules and three open-book FCIA examinations. The five tracks are:

Finance, Investment, and ERM

Individual Life Insurance and Annuities

Group Benefits

Property and Casualty Insurance

Retirement Benefits

China

China Association of Actuaries (CAA) was established in November 2007. CAA promotes the actuarial profession in China, administers accreditation examinations toward the designations of associateship and fellowship, and organizes its members' continuing professional education.

There are eight examinations required for associateship of CAA: mathematics, financial mathematics, actuarial modeling, economics, actuarial mathematics of life insurance, actuarial mathematics of nonlife insurance, accounting and finance, and actuarial management. Actuarial science programs at universities in China train students in examination subjects required for associateship. Applicants who have passed exams of SOA, CAS, or IFoA can be exempted from at most four examinations towards associateship of CAA.

At the fellowship level, CAA offers two tracks of specialization: life insurance and nonlife insurance, each requiring five examinations. In addition to passing the five required examinations for either track, an associate who wishes to become a fellow must have three or more years of experience in actuarial or financial functions and secure two fellows' recommendations.

China – Hong Kong

The Actuarial Society of Hong Kong (ASHK) conducts its own local actuarial examination (on top of the overseas professional examinations) applicable to new Fellow of the ASHK. Typically, to be admitted as a Fellow of the ASHK, the member must also be a Fellow of one of the actuarial bodies of Australia, Canada, the United Kingdom, or the United States, although there is an increasing number from other countries, especially European ones. Under the Hong Kong government's insurance company (qualification of actuaries) regulations, the qualifications for appointment as an Appointed Actuary are Fellow of the Institute and Faculty of Actuaries in the United Kingdom, Fellow of the Institute of Actuaries of Australia, or Fellow of the Society of Actuaries of the United States of America. Students who wish to pursue an actuarial science degree can do so at six local universities—the Chinese University of Hong Kong, Hang Seng University of Hong Kong, Hong Kong Polytechnic

University, the University of Hong Kong, the Hong Kong University of Science and Technology, and City University of Hong Kong—that all offer a range of actuarial subjects.

Denmark

The Danish Society of Actuaries was established in 1901. It works closely with the AAE and its main objective is the advancement of actuarial science and to promote the interests of the actuarial profession in Denmark. The society participates in all national hearings on actuarial concerns and is often represented on government-appointed committees.

Actuaries in Denmark are usually divided into nonlife insurance actuaries and life insurance actuaries. All life insurance companies and pension funds must employ a *responsible actuary* approved by the Danish Financial Supervisory Authority (DFSA). Furthermore, the DFSA approves the appointment of key function holders, including the actuarial function holder, in both life and nonlife insurance companies. Actuarial education in Denmark is university-based. Most actuaries in Denmark have a master's degree in actuarial mathematics from the University of Copenhagen. To become a *responsible actuary*, you must have a Danish Master's degree in actuarial mathematics or similar qualifications and, in addition, at least 5 years of actuarial experience, of which 1 year must be in close collaboration with a responsible actuary.

Actuarial theory and practice in Denmark are closely linked, among other reasons, because multiple scholars from the Section of Insurance and Economics at the Department of Mathematical Sciences of the University of Copenhagen are active members of the Danish Society of Actuaries.

Finland

The actuarial profession in Finland is organized through the Actuarial Society of Finland, which has approximately 300 members. About one-third of them are fully certified actuaries. However, the government formally controls actuarial education and actuarial accreditation. The Ministry of Social Affairs and Health of Finland nominates an Actuarial Examination Board, which administers relevant examinations and controls the syllabus and the qualification standards. The Ministry works closely with the Actuarial Society of Finland in the sense that the members of the Examination Board, for example, are usually also members of the Actuarial Society. However, the Insurance Supervisory Authority has additional resources for developing actuarial education and research, upon which the Board also draws.

Admission to the Fellowship of the Actuarial Society of Finland is granted upon successful completion of a relevant university degree, the completion of actuarial foundation courses, the passing of additional examinations dealing with actuarial applications, the writing of a thesis, and the completion of at least 1 year of practical actuarial work. Universities offer the foundation courses, whereas the Examination Board prepares the actuarial application examinations. Foundation courses cover risk mathematics, survival models, financial mathematics, and basic life insurance. Many of the examinations are written while the candidates are fully employed, so it usually takes several years before they are able to qualify for Fellowship.

To enter the actuarial profession, graduates must have a Master's degree with a major in mathematics or cognate discipline, provided that a sufficiently high standard in mathematics has been demonstrated. Courses in probability, statistics, and stochastic processes are particularly relevant. Only some universities offer actuarial foundation courses. The University of Helsinki, on the other hand, also has an M.Sc. program in mathematics, with specialization in actuarial studies. In order to implement the AAE Core Syllabus, changes have been made in the education system. Courses on financial economics and investment mathematics have become mandatory and a course in economics has been added to the syllabus. The examinations in actuarial applications include four general examinations and an individualized self-study test. At the general level, the subjects of the tests are insurance legislation, insurance accounting, and applied insurance mathematics. The applied insurance mathematics examination covers actuarial modeling, practical risk theory, solvency issues, and investments.

At the specialized level, candidates select one of the following subjects: life insurance, mandatory pensions, or general insurance. Along with general principles and practice, Finnish conditions are emphasized. This holds, in particular, for pensions because of Finland's unique mandatory pension system. The Examination Board supervises the thesis. Starting in 2005, two options became available. First, candidates are able to write a brief analysis of an issue of actuarial concern. The purpose of this type of thesis is to demonstrate the ability to present ideas and arguments. The second option is to write a research paper on a practical topic. Many candidates have a great deal of work experience before attaining full professional status. The aim of this type of thesis is therefore to encourage the development of that experience and foster innovations in actuarial science. The Foundation for Promotion of the Actuarial Profession actually encourages this type of work by providing financial support. Finally, candidates must have completed at least three years of practical experience in an insurance company or have done equivalent work. Experience may count as equivalent if it consists of practical applications of actuarial methods, under the supervision of a Fellow of the Society of Actuaries. Continuing professional development is not mandatory.

France

The Institute of Actuaries (French: Institut des Actuaires) is the association of actuaries in France and now has more than 5000 members. The Institute was created in 2001 by a merger of the Institute of Actuaries of France and the French Federation of Actuaries. The Institute is a full member of the International Actuarial Association and the AAE. Responsible for organizing and representing the actuarial profession in France, the Institute of Actuaries is the guarantor of compliance with the standards and professional ethics of its members.

The education of actuaries in France is university based. Eight universities offer degree programs in actuarial science: College of Engineers, National School of Statistics and Economics Administration (ENSAE Paris Tech), ESSEC Business School, The Euro-Institute of Actuarial Science of the University of Brest, Institute of Financial Science and Insurance, Institute of Statistics of University of Paris, Paris Dauphine University, and University of Strasbourg. Professionals can also become actuaries through the National Conservatory of Arts and Crafts (Conservatoire national des arts et métiers) or through the Certificate of Actuarial Expertise (Certificat d'Expertise Actuarielle). The Institute of Actuaries accredits the university programs and issues the title of actuary. A candidate is generally accepted as an associate actuary who then must demonstrate competences in order to be recognized as qualified actuary, member of the Institute of Actuaries.

Germany

German Association of Actuaries (Deutsche Aktuarvereinigung in German, DAV in short) is a full member of AAE. German Society for Insurance and Financial Mathematics (Deutsche Gesellschaft für Versicherungs- und Finanzmathematik e.V., DGVFM) is the scientific partner organization of DAV, and the Institute of Pension Actuaries ("Institut der Versicherungsmathematischen Sachverständigen für Altersversorgung e.V.," IVS) is a branch of DAV focusing on occupational pensions. In 2000, the three organizations founded German Actuarial Academy (DAA) to guarantee high-quality education and training for German actuaries. Education and accreditation are split up between DAV and DAA. While the Academy (DAA) coordinates all training events, the exams are conducted and assessed by the DAV.

A completed program of mathematics at a university in Germany is required to begin the training toward "Actuary DAV." This can be substituted by a diploma in physics, or a diploma in statistics, as well as the First State Examination for Secondary Level II in Mathematics. This prerequisite can also be satisfied with 90 or more credit points in mathematics through European Credit Transfer System (ECTS), or by passing an admission exam in mathematics. Another prerequisite before the training for Actuary DAV is basic knowledge of stochastics, of which the objectives are prescribed by the Education Committee. Candidates satisfy this requirement by taking university courses or having 30 credit points in accordance with ECTS.

To qualify as an Actuary DAV, candidates must pass 10 examinations and take two courses. The DAV examination system is divided into six subjects of basic knowledge and four subjects of specialization knowledge. The examination subjects of basic knowledge are Economic and Legal Environment, Applied Statistics, Financial Mathematics and Risk Assessment, Insurance Mathematics, Modeling and ERM, and Corporate Management. Additionally the mandatory courses in "Communications" and "Professionalism" need to be completed at this stage. Exemptions for the examinations in the area of basic knowledge are allowed by passing equivalent exams administered at university. University professors can apply at DAV to become professors of trust, who are permitted to certify that a university course is equivalent to a DAV examination.

At the next stage, candidates must choose four supplementary examinations in the area of specialist knowledge in order to complete the education as An actuary DAV. Each candidate must pass two exams on a specialization subject chosen from eight options: life insurance, health insurance, pensions, casualty insurance, finance, building society savings, actuarial data science, and enterprise risk management. The remaining two examinations can be on any one or two specialization subjects that are different from the candidate's chosen specialization field. No exemptions of specialist knowledge exams are allowed for any university courses.

To become a member of DAV 3 years of actuarial work experience are necessary, of which at least 2 years must have been gained in the selected area of special knowledge.

India

Actuarial education in India is profession-based. The Institute of Actuaries of India (IAI) was constituted in 2006 by an act of Parliament, after the dissolvement of its predecessor Actuarial Society of India, established in the year 1944. IAI offers a series of examinations that candidates must pass to qualify as an actuary.

Candidates who have passed all seven Core Principles series (one exam on each of three subjects—Business Finance, Business Economics, and Business Management—and two papers, each for the remaining four subjects, viz., Actuarial Statistics, Risk Modeling and Survival Analysis, Actuarial Mathematics, Financial Engineering and Loss Reserving) and Core Practices series (one exam on each of three subjects—Actuarial Practice, Modeling Practice, and Communications Practice) can become Associate of IAI (AIAI) upon application.

The designation of Fellowship can be attained by passing any two subjects from the Specialist Principle series and one subject from Specialist Advanced series. The specialist principle series comprises seven subjects (Health and Care, Life Insurance, Pension and Other Benefits, Investment and Finance, Financial Derivatives, General Insurance Reserving and Capital Modeling, and General Insurance Pricing) and specialist advanced series offers five subjects (Health and Care, Life Insurance, Pension and Other Benefits, Investment and Finance, and General Insurance).

Ireland

The Irish equivalent of a Fellow of the Society of Actuaries is a Fellow of the Society of Actuaries in Ireland (FSAI). Most Fellows qualify through the Institute and Faculty of Actuaries in the United Kingdom. Under the rules of the Society of Actuaries in Ireland (SAI), all Fellows must either (1) have met the requirements of Fellows of the Institute and Faculty of Actuaries of the United Kingdom (IFoA) or (2) be fellows of actuarial associations that maintain an MRA with SAI, including members of AAE, SOA, AIA, CIA, and CAS, providing they meet experience requirements and complete professionalism training prescribed by the SAI. At present, associates of IFoA and the Actuarial Society of South Africa can become Associates of SAI upon application.

Israel

The education of actuaries in Israel is concentrated in universities. In addition to courses in actuarial science available at the Hebrew University in Jerusalem and at Tel Aviv University, the University of Haifa maintains an active research center in actuarial science, offering a Master's degree. The Israel Association of Actuaries is the professional body for actuaries in Israel and is a full member of the International Actuarial Association. It has a mutual recognition agreement with the Institute and Faculty of Actuaries. One of its functions is to enhance the practical knowledge of graduates of the academic courses in Israel and abroad and examine these candidates for Fellowship.

The actuarial training system in Israel is coordinated with the training system of the Institute and Faculty of Actuaries and emphasizes three main components: examinations, professionalism, and practical work experience. Memberships are in two categories: specialized actuary (Associate) and full actuary (Fellow).

Italy

Actuarial profession in Italy is regulated and protected by the law, and the governance is in the hands of "Consiglio Nazionale degll Attuari" (CNA). It is a public body responsible for managing, driving, and monitoring all activities of the actuarial profession, including the international representation in the Actuarial Association of Europe (AAE) and in the International Actuarial Association (IAA), through the Italian Society of Actuaries (ISOA).

The ISOA is a member of the AAE and therefore has reciprocal agreements with the member countries of that group. The CNA maintains a permanent professional development program through the Actuarial Training Courses Program (CPD program), compulsory by the law, allowing its members to keep up to date with changes in actuarial practice resulting from globalization and European integration.

The actuarial education in Italy is university-based and the title of fully qualified actuary is obtained through a state examination. A Master's degree in finance, actuarial science, or statistics is required to register for state examination of professional actuary. Furthermore, admission to the state examination for junior actuary requires a 3-year degree program in statistics.

Japan

The actuarial education in Japan is profession-based. The Institute of Actuaries of Japan (IAJ) offers actuarial courses that enable applicants to acquire basic knowledge and to prepare for qualification examinations.

Actuarial courses are divided into two categories, basic and advanced courses. The basic courses are intended for students of the Institute, while advanced courses are aimed at people who have completed the basic subjects. To become an Associate member of the Institute, candidates must pass examinations in the following five basic courses:

Mathematics
Life insurance mathematics
Nonlife insurance mathematics
Pension mathematics
Accounting, economics, and investment theory.

After passing these courses, candidates qualify for Associate membership in the IAJ. To become a Fellow of the Institute (in Japan an actuary is a person who is a Fellow of the IAJ), Associates must pass two additional advanced courses: (LI) Life Insurance I and II, or (NLI) Nonlife Insurance I and II, or (P) Pension I and II. Professionalism training is also part of the fellowship requirements.

Another characteristic of the IAJ credential is that it is based on the premise of "working practically as an actuary." Most Fellows earn their certification by studying and taking the exams while engaging in actuary work at their organization. One of the qualification requirements for some positions, such as "Hoken Keirinin (Appointed Actuary)" and "Nenkin Sūrinin (Certified Pension Actuary)," is fulfilled by being an IAJ Fellow.

Malaysia

The actuarial profession in Malaysia is represented by the Actuarial Society of Malaysia. Actuaries who are Fellows of the following institutions may be admitted as Fellows of the Society: the Institute and Faculty of Actuaries of the United Kingdom, the Society of Actuaries of America, the Casualty Actuarial Society, or the Institute of Actuaries of Australia. Admission to the Society must be approved by the Executive Committee of the Society. Qualified actuaries are allowed to practice in Malaysia if they reside in Malaysia or, in the opinion of the Executive Committee, are familiar with Malaysian conditions, and have paid the requisite admission and annual membership dues. Fellows of the Society can become Appointed Actuaries of insurance companies by being approved by the regulatory authority in Malaysia (Bank Negra Malaysia). Appointed Actuaries must be residents of Malaysia and have at least 1 year of relevant work experience with a Malaysian insurer.

Mexico

The education of actuaries in Mexico is university-based. To be able to work as an actuary in Mexico, and to be allowed to use the designation "actuary," candidates must fulfill three requirements. (1) They must complete a 4-year undergraduate program in actuarial science, which includes 480 hours of unpaid socially valuable work. The Mexican syllabus is close to that prescribed by the SOA. In fact, many students in Mexico are encouraged to write the SOA examinations. (2) They must write a relevant dissertation. (3) They must defend the dissertation before an examination committee.

Some universities accept graduate work in relevant academic programs and the passing of written and oral comprehensive examinations as dissertation equivalents. In order to be accredited as actuaries with signing privileges, graduates must have their university degrees approved by the Ministry of Education and obtained from the Ministry of Professional license. Certified actuaries are publicly sworn to uphold the code of ethics of the profession, but they are not required to become members of the Colegio or any other association of actuaries. A significant number of actuaries in Mexico work in nontraditional areas such as finance, government, planning, and information technology.

Mexico has two actuarial organizations: In 1962, the Asociacion Mexicana de Actuarios del Seguro de Vida was formed. Its members tend to work in life insurance. In 1980, the association expanded its membership to include all actuaries and became the Asociacion Mexicana de Actuarios. In addition, the profession established the College of Actuaries in 1867, the Colegio de Actuarios de Mexico, which was transformed into the Colegio Nacional de Actuarios in 1982. Membership in the College is not required to function as an actuary.

Netherlands

The Dutch equivalent of Fellowship in the Society of Actuaries is a Fellowship in the Actuarieel Genootschap (Royal Actuarial Society). Three roads lead to this Fellowship:

1. Successful completion of the actuarial program of the Actuarieel Instituut (Actuarial Institute). The Actuarial Institute offers an Executive Master of Actuarial Science program that runs five courses and six case studies. The courses are statistical methods, life and pensions, valuation and hedging,

risk and regulation, and capita selecta actuarial science. A thesis is mandatory at the end of a 2-year period of the program. Admission requirements include a strong background in probability, calculus, linear algebra, and statistics, knowledge of risk theory and life insurance, and knowledge of economics and quantitative finance. Premaster program offers modules in required subjects to help the applicants in need.

2. The other option is to complete a Master's program in actuarial science at the University of Amsterdam. The completion of the program requires 60 credits from 6 general courses, 3 track-specific electives, and a thesis. The general courses are asset liability management—cases and skills, financial mathematics for insurance, nonlife insurance—statistical techniques and data analysis, principles of mathematics and economics of risk, risk management for insurance and pensions, and stochastic calculus. Upon finishing the program, graduates can participate in the Post-Master's Actuarial Practice Cycle, success in which qualifies students for memberships of the Royal Actuarial Society.

3. Fellows working in Netherlands are members of associations that maintain a mutual recognition agreement with Royal Actuarial Society.

The Netherlands is a member of the Actuarial Association of Europe, which allows for recognition of the Fellowships of other member countries.

Norway

The education of actuaries in Norway takes place primarily at universities. Both the University of Oslo and the University of Bergen have degree programs of study that provide actuarial competence to students and qualify graduates to be admitted as members of the Norwegian Actuarial Association.

Since 1916, the University of Oslo has offered a degree program in actuarial science in insurance mathematics and statistics. In the 1950s and 1960s, a program of actuarial studies based on stochastic principles was established. Risk theory and nonlife insurance were added to the syllabus in the 1970s. Now the University of Oslo offers a Finance, Risk, and Insurance option to students of the Master's Degree program Stochastic Modelling, Statistics and Risk Analysis. Since 1997, the University of Bergen has also offered a degree program in actuarial science. Currently the University of Bergen runs a 5-year integrated program in Actuarial Science and Data Analytics from which students receive Master's degrees.

Portugal

The Portuguese equivalent of a Fellow of the Society of Actuaries is an Actuário Titular, a full member of the Portuguese Institute of Actuaries (IAP).

The education of actuaries is university-based. The Universidade Técnica de Lisboa (Technical University of Lisbon) and the Universidade Nova de

Lisboa (New University of Lisbon) offer the two master programs in actuarial science that comply with the Syllabus of AAE.

To be awarded the title of Fellow of the IAP, applicants must have a university curriculum containing subjects for basic actuarial training, completed by specific training in the actuarial area, in other words, a university curriculum that globally satisfies the actuarial contents contained in the IAP Core Syllabus —same as the AAE Core Syllabus. Recent experience is also required. Specifically, an applicant must have working experience as an actuary 50% of the time in the last 10 years.

Singapore

The Singapore Actuarial Society (SAS) aims to be the recognized representative body of the actuarial profession in Singapore, having the final authority in setting professional standards.

The fellowship and associateship of SAS can be earned by members of other actuarial associations.

Fellows of one of five recognized actuarial associations, namely, IFoA, SOA, Actuaries Institute Australia, Canadian Institute of Actuaries, and CAS, are recognized as fellows of SAS upon application. Fellows of full-member associations of the International Actuarial Association, other than the five named above, can become fellows of SAS after assessment by the Council of SAS. The assessment examines the applicant's professional actuarial qualifications, practical experience, and other criteria deemed necessary.

Associates from the five recognized actuarial associations and fellows from full-member associations of the IAA except the five recognized associations can become associates of SAS.

South Africa

Starting in late 2006, the Actuarial Society of South Africa (ASSA) formally embarked on establishing a local actuarial education system. The intention is for the education system to rely heavily on the university system. The mechanism through which ASSA engages with the university system is via granting exemption from ASSA subjects to students who pass subjects at a specified level at accredited universities. The accreditation process is designed to encourage individual strengths within the universities and local variation in actuarial courses while ensuring that ASSA's requirements to approve exemption are met.

Accredited universities include Cape Town University, Pretoria University, Johannesburg University, Stellenbosch University, Witwatersrand University, Free State University, Northwest University, and Nairobl University, with varying accredited subjects.

Since 2010, ASSA has offered its own Fellowship (FASSA) and Associateship designations as internationally recognized professional qualifications. ASSA enters mutual recognition agreement with Actuarial Association of

Europe, CAS, the Canadian Institute of Actuaries, Deutsche Aktuarvereinigung (Germany), IFoA, the Institute of Actuaries of Australia, Israel Association of Actuaries, the Netherlands Koninklijk Actuarieel Genootschap, and Society of Actuaries Ireland.

South Korea

The following is the response from ChatGPT on February 26, 2023, to the question, "What are the requirements for becoming an actuary in South Korea?"

To become an actuary in South Korea, you need to fulfill the following requirements:

Educational requirements: You must have a bachelor's degree in actuarial science, mathematics, statistics, or another related field from an accredited university.

Actuarial Examinations: Pass the required actuarial exams set by the Korean Institute of Actuaries (KIA) in areas such as probability, financial mathematics, financial economics, life insurance, and more.

Experience: You must have at least three years of relevant work experience in actuarial science, preferably in insurance or finance.

KIA membership: You must become a member of the Korean Institute of Actuaries (KIA) by passing the KIA's membership examination and fulfilling the other membership requirements.

The KIA requires candidates to pass the following actuarial exams to become an actuary in South Korea:

Probability Exam
Financial Mathematics Exam
Financial Economics Exam
Life Insurance Exam
Nonlife Insurance Exam
Pension and Employee Benefits Exam
Enterprise Risk Management Exam

Spain

Actuarial profession is a regulated profession in Spain, established by law. Since the establishment of the Instituto de Actuarios Españoles (Institute of Spanish Actuaries), the Spanish professional association, the only requirement to be admitted as a full member is to have the actuarial or financial sciences master's degree plus compliance with the Code of Conduct.

The higher education degree, approved by Institute of Spanish Actuaries and offered by nine universities through their faculties of economics and business administration, is a full actuarial education program that takes 2 years and consists of 120 credits (one credit involves 20 effective lecture hours), following the Core Syllabus of the Actuarial Association of Europe, which has been approved by the Institute of Spanish Actuaries.

Almost half of the credits must be in the following subjects: actuarial statistics (including topics on stochastic processes, survival models, and, partially, risk theory), financial mathematics (also including topics on investment), actuarial mathematics (including risk theory and life and nonlife insurance mathematics), accounting and financial reporting in insurance, banking and investment, insurance, banking and stock market regulations, and social security economics and techniques. In deciding on the rest of the program or syllabus, each university has a significant degree of autonomy, but most of them expand the number of credits in actuarial mathematics, financial mathematics, statistics, accounting, and financial reporting and then offer specialized courses in private pension plans, financial instruments and markets, taxation, solvency, reinsurance, insurance and financial marketing, computing, and so on. Students who want to pursue such a program must have an undergraduate degree, usually in economics or business administration. It must include courses in mathematics, probability and statistics, economics, finance and accounting, and financial reporting.

In addition, the Institute of Spanish Actuaries has implemented the changes required to meet the education requirements set by the Actuarial Association of Europe in its new Core Syllabus.

Sweden

Swedish Actuarial Association (Svenska Aktuarieforeningen) is a full member of the AAE. The Swedish equivalence of any AAE member association's fellow is called Certified Member. Certified membership in the Svenska Aktuarieforeningen is granted to individuals who have fulfilled appropriate academic requirements. (1) Candidates must hold a Master's degree in actuarial mathematics at a Swedish university or equivalent. (2) Candidates must also have completed courses that cover the subjects listed below:

Basic mathematics
Basic mathematical statistics
Life insurance mathematics
Nonlife insurance mathematics
Financial mathematics
Insurance accounting
Economics
Insurance law

Other requirements address language (English and Swedish) competency and actuarial work experience.

Switzerland

In Switzerland, the profession is made up, as everywhere else, of actuaries with university degrees who have passed special professional examinations and have special legislative powers assigned to them in their capacity as general insurance and life insurance actuaries.

Swiss actuaries receive their qualification from the Swiss Association of Actuaries and hold the title "Actuary SAA." Actuarial studies in actuarial mathematics in Switzerland involve completing a course of studies based on the Swiss syllabus. Swiss universities of ETH Zurich, Berne, Lausanne, and Basel are accredited by the SAA. An appropriate degree from these universities means that the academic requirements have been met. In addition, actuaries must have three years of practical experience and have passed the examination colloquium.

Pensions in Switzerland are subject to special laws and pension contributions are mandatory for all employers. Pension Actuary in Switzerland is a Federal Diploma approved by the Swiss Secretariat for Education (SBFI) and supervised by the OAK-BV (Supervision Commission for Occupational Pensions). As a result, pension actuaries must obtain additional qualifications to practice. The Swiss Association of Actuaries offers and steers the education as well as controls the quality of the lectures and exams to get the Federal diploma.

United Kingdom

The Institute and Faculty of Actuaries was created in 2010 by merging the Institute of Actuaries in England and the Faculty of Actuaries in Scotland to represent actuaries in the United Kingdom. This association maintains and manages an examination and accreditation system not too different in content from that of the Society of Actuaries in the United States, with strong emphasis on professionalism and work experience.

- Associateship of IFoA
  There are two layers of exam requirements for associateship of IFoA: Core Principles and Core Practices.

  Core Principles are made up of three modules: Actuarial Statistics, Actuarial Mathematics, and Business.

  Exams CS1—Actuarial Statistics and CS2—Risk Modeling and Survival Analysis are the building blocks for the Actuarial Statistics Module.

  Exams CM1—Actuarial Mathematics and CM2—Financial Engineering and Loss Reserving are the building blocks for the Actuarial Mathematics Module.

  The Business Module encompasses three exams: CB1—Business Finance, CB2—Business Economics, and CB3—Business Management.

  Core Practices include three modules: Actuarial Practice, Modeling Practice, and Communication Practice. Actuarial Practice (CP1) builds on the technical material covered in all the exams mentioned earlier and uses the techniques learned to solve practical problems that might arise in any area in which actuaries practice. Modeling Practice (CP2) builds on material covered in earlier subjects and seeks to equip a student with more "rounded" business skills and the prime emphasis is on good communication when using and presenting spreadsheet work. Both CP1 and CP2 require students to complete two papers (written answer exams). Communication Module (CP3) is a single written exam that assesses examinees' ability to communicate effectively in writing to a nonactuarial audience.

To obtain Associate status of IFoA, an individual also must have 2 years of practical experience and complete Stage 1 and 2 Online Professionalism Courses.

- Fellowship of IFoA

On top of the Associate requirements, the fellowship requirements include two Specialist Principles (SP) exams, one Specialist Advanced (SA) exam, practical work experience called Personal and Professional Development, and Professional Skills Course.

Subjects of Specialist Principles (SP):

SP1: Health and Care
SP2: Life Insurance
SP4: Pensions and Other Benefits
SP5: Investment and Finance
SP6: Financial Derivatives
SP7: General Insurance Reserving and Capital Modeling
SP8: General Insurance Pricing
SP9: Enterprise risk management
SP10: Banking

There is no SP3.
Subjects of Specialist Advanced (SP):

SA1: Health and Care
SA2: Life Insurance
SA4: Pensions and Other Benefits
SA7: Investment and Finance
SA3: General Insurance
SA10: Banking

All examinations at both associate and fellow levels can be exempted, usually by taking specified college courses from approved colleges or universities. An alternative to Specialist Advanced subjects is a research dissertation from a Master's degree program.

The IFoA has reciprocity agreements with actuarial associations of Australia, India, Israel, and South Africa and the countries belonging to Actuarial Association of Europe.

Other countries

The list of countries covered in this section is obviously incomplete. Most countries around the world have insurance companies and either privately run or public pension schemes. Actuarial considerations are therefore relevant to all countries. Rather than being encyclopedic, the choice of countries profiled in this section is intended to give you an idea of the variety of different national traditions and models for being an actuary. It also sheds some light on international mobility.

In her article on the actuarial education in Arab countries, Dana Barhoumeh (see Barhoumeh (2018)) notes that most actuarial work within the region is undertaken by actuaries from outside, mostly from South Asia and Europe. One of the reasons for the shortage of local actuaries has been the historical lack of recognition of the profession. On the other hand, in other countries' actuarial associations, one can find actuaries who are originally from the Middle East. In fact, the Gulf Actuarial Society was set up to bring together IFoA members based in the six Gulf states of Bahrain, Kuwait, Oman, Qatar, SaudI Arabia, and the United Arab Emirates (UAE), to raise the profile of the actuarial profession in the area, among other goals.

Chapter 2

Actuarial education

All actuarial societies, be it the Society of Actuaries, the Canadian Institute of Actuaries, the Casualty Actuarial Society, the Institute and Faculty of Actuaries, the Actuarial Association of Europe, and others, expect their members to have studied certain subjects and passed a number of examinations in core courses. In this section, you will get a glimpse of what knowledge is required.

2.1 The International Syllabus

The International Actuarial Association (IAA) publishes the IAA Education Guidelines and IAA Education Syllabus. It identifies a number of academic topics as describing the repertoire of scientific knowledge and competency areas of actuaries. The Guideline and Syllabus summarize the understanding of leading actuaries of the core tools of their profession. IAA requires full-member associations to have education requirements that are at least equivalent to the IAA Syllabus. Here is the list of the current version of the Syllabus, which divides actuarial education into nine broad areas:

Statistics

Aim: To enable students to apply core statistical techniques to actuarial applications in insurance, pensions, and emerging areas of actuarial practice.

Topics: random variables; regression; Bayesian statistics and credibility theory; stochastic processes and time series; and simulation.

Economics

Aim: To enable students to apply the core principles of microeconomics, macroeconomics, and financial economics to actuarial work.

Topics: Microeconomics; macroeconomics; financial economics.

Actuaries' Survival Guide, Third Edition. DOI: 10.1016/B978-0-443-15497-3.00002-4
© 2025 Elsevier Inc. All rights reserved, including those for text and data mining, AI training, and similar technologies

Finance

Aim: To enable students to apply the core principles of financial theory, accounting, corporate finance, and financial mathematics to actuarial work.

Topics: Financial reporting and taxation; securities and other forms of corporate finance; financial mathematics; corporate finance.

Financial systems

Aim: To enable students to understand the financial environment in which most actuarial work is undertaken, and key products and principles of insurance, pensions, and other areas of traditional and emerging actuarial practice.

Topics: Role and structure of financial systems; participants in financial systems; financial products and benefits; factors affecting financial system development and stability.

Assets

Aim: To enable students to apply asset valuation techniques and investment theory to actuarial work.

Topics: Investments and markets; asset valuation; portfolio management; investment strategy and performance measurement.

Data and systems

Aim: To enable students to apply methods from statistics and computer science to real-world data sets in order to answer business and other questions, in particular with application to questions in long- and short-term insurance, social security, retirement benefits, healthcare, and investment.

Topics: Data as a resource for problem-solving; data analysis; statistical learning; professional and risk management issues; visualizing data and reporting.

Actuarial models

Aim: To enable students to apply stochastic processes and actuarial models to actuarial work, in particular to applications in long- and short-term insurance, social security, retirement benefits, healthcare, and investment.

Topics: Principles of actuarial modeling; fundamentals of severity models; fundamentals of frequency models; fundamentals of aggregate models; survival models; actuarial applications.

Actuarial risk management

Aim: To enable students to apply core aspects of individual risk management and enterprise risk management to the analysis of risk management issues faced by an entity, and to recommend appropriate solutions.

Topics: The risk environment; risk identification; risk measurement and modeling; risk mitigation and management; risk monitoring and communication.

Personal and actuarial professional practice

Aim: To enable students to apply their technical knowledge and skills in an effective, practical, and professional manner.

Topics: Effective communications; problem-solving and decision-making; professional standards; professionalism in practice.

Mathematics topics fundamental to actuarial study are prerequisites, as mastering such topics enables students to develop an adequate foundation upon which to build the additional mathematical skills required for successful actuarial practice. Fundamental mathematics topics are functions and sets, differentiation, integration, sequences and series, differential equations, real and complex numbers, matrices and systems of linear equations, vectors, vector spaces and inner product spaces, and probabilities.

This list can be found in the Education Syllabus section of the website of the International Actuarial Association. It is a wonderful conceptual organizer for the overwhelming mass of mathematical, economic, financial, and other ideas that make up the syllabus upon which the SOA and CAS examinations are based. As you read on, you might try to fit the listed topics and sample examination questions into this scheme. It will help you with the conceptual order and organization of the material that follows.

2.2   The European Syllabus

The Actuarial Association of Europe (formerly Groupe Consultatif Actuariel Europeen) was established in 1978 to bring together the actuarial associations in the European Union to represent the actuarial profession in discussions with European Union institutions on existing and proposed EU legislation that has an impact on the profession. There are two types of AAE memberships: full members that meet certain professional criteria and observer members.

Full members include actuarial associations in most European countries – Austria, Belgium (two associations), Bulgaria, Channel Islands, Croatia, Cyprus, Czech Republic, Denmark, Estonia, Finland, France, Germany, Greece, Hungary, Iceland, Ireland, Italy, Latvia, Lithuania, Luxembourg, Netherlands, Norway, Poland, Portugal, Romania, Slovakia, Slovenia, Spain (two associations), Sweden, Switzerland, Turkey, and the UK. There are four observer members currently – actuarial association of Malta, Montenegro, Serbia, and Ukraine. The AAE currently represents over 27,000 actuaries.

By design, the AAE Core Syllabus covers all topics on the IAA Education Syllabus as basic actuarial education. In addition, it also contains a section on advanced skills required of actuaries who specialize in different areas of actuarial practice. The specialization areas available include life, pensions, general insurance, enterprise risk management, investments, health care, accounting, banking, social security, reinsurance, management and leadership, and data science. Recognizing that some learning objectives might not be achieved only through theoretical studies, the AAE Core Syllabus stipulates "a minimum of two years' of practical actuarial experience should be required by all Full Membership Associations to fulfill the requirements."

The syllabus also deals with the postqualification training necessary to ensure that actuaries are up to date with changes in the framework for their practice area and includes a continuing professional development (CPD) component.

Even a cursory reading of the SOA, CAS, and AAE syllabi shows that there is remarkable agreement among actuaries in the countries concerned regarding what academic and profession skills actuaries should have and should continue to develop for professional competency and lifelong success.

The details of the AAE's syllabus are available on the AAE's website.

2.3   The North American Syllabus

There is an old saying: The more things change, the more they stay the same. During the first quarter of the century, the organization of the actuarial examinations and distribution of topics over the underlying courses have changed. Some topics were reorganized, some examinations are now online or administered as computer-based tests, and professional and experiential components have been strengthened. Nevertheless, the mathematical underpinnings of the profession are essentially the same, now and then. However, the core of this book consists of the wisdom and experience expressed in the results of a survey, which readers found both stimulating and informative.

As presented in the "Actuaries around the World" section in Chapter 2, there are major differences in actuarial education around the world. However, the SOA and CAS qualifications are respected and honored across the globe. In the United States, Canada, and many countries, SOA exams are now administered through Prometric test centers, and contact information can be found on the Prometric website. If candidates need to make special arrangements, they should contact SOA directly at soaexams@soa.org. The CAS-administered examinations utilize computer-based testing (CBT) in partnership with Pearson VUE Testing Centers. Both exam service companies operate exam centers in many countries around the world to accommodate those who pursue actuarial careers. In Canada, the CIA introduced a new pathway to Associateship recently.

2.4   The SOA Examinations

The Society of Actuaries (SOA) currently offers three pathways to professional accreditation: Associateship (ASA), Fellowship (FSA), and Chartered Enterprise Risk Analyst (CERA). The website of the Society provides an interactive survey of the requirements.

Associateship requirements

According to the Society, an Associate of the SOA has demonstrated knowledge of the fundamental concepts and techniques for modeling and managing risk. The Associate has also learned the basic methods of applying those concepts and techniques to common problems involving uncertain future events, especially those with financial implications. Candidates for Associateship are required to demonstrate that they possess this knowledge by passing a number of required courses. The requirements can be classified into four parts: six examinations, VEE (validation by education experience) subjects, e-learning courses, and a seminar on ethics and professionalism. The Society recommends a typical order to accomplish all requirements.

Examinations

There are six exams that students have to pass before they earn the Associateship from the Society. The exams are Financial mathematics (FM), Probability (P), Fundamentals of Actuarial Mathematics (FAM), Statistics for Risk Modeling (SRM), a choice of Advanced Short Term Actuarial Mathematics (ASTAM) or Advanced Long Term Actuarial Mathematics (ALTAM), and Predictive Analytics (PA). The SOA website describes in detail the examination and accreditation process.

- Exam FM (Financial Mathematics): According to the Society, the syllabus for Exam FM develops the candidate's understanding of the fundamental concepts of financial mathematics, and how those concepts are applied in calculating present and accumulated values for various streams of cash flows as a basis for future use in reserving, valuation, pricing, asset/liability management, investment income, capital budgeting, and valuing contingent cash flows. A basic knowledge of calculus and an introductory knowledge of probability is assumed.
- Exam P (Probability): The syllabus for Exam P develops the candidate's knowledge of the fundamental probability tools for quantitatively assessing risk. The application of these tools to problems encountered in actuarial science is emphasized. A thorough command of the supporting calculus is assumed. Additionally, a very basic knowledge of insurance and risk management is assumed.

- Exam FAM (Fundamentals of Actuarial Mathematics): The syllabus for the short-term section of the examination provides an introduction to modeling and covers important actuarial methods that are useful in modeling. It will also introduce the candidate to the foundational principles of ratemaking and reserving for short-term coverage. The syllabus for the long-term section of the examination develops the candidate's knowledge of the theoretical basis of contingent payment models and the application of those models to insurance and other financial risks. A thorough knowledge of calculus, probability (as covered in Exam P), mathematical statistics (as covered in VEE Mathematical Statistics), and interest theory (as covered in Exam FM) is assumed.

- Exam SRM (Statistics for Risk Modeling): The syllabus for Exam SRM provides an introduction to methods and models for analyzing data. Candidates will be familiar with regression models (including the generalized linear model), time series models, principal components analysis, decision trees, and cluster analysis. Candidates will also be able to apply methods for selecting and validating models. A thorough knowledge of calculus, probability (as covered in Exam P), and mathematical statistics (as covered in VEE Mathematical Statistics) is assumed.

- Exam ASTAM (Advanced short-term actuarial mathematics): The syllabus for Exam ASTAM continues to develop the candidate's knowledge of modeling and important actuarial methods that are useful in modeling, as well as ratemaking and reserving for short-term coverages. A thorough knowledge of calculus, probability (as covered in Exam P), mathematical statistics (as covered in VEE Mathematical Statistics), and the fundamentals of short-term actuarial mathematics (as covered in FAM) is assumed. This is a written-answer exam.

- EXAM ALTAM (Advanced long-term actuarial mathematics): The syllabus for Exam ALTAM continues to develop the candidate's knowledge of the theoretical basis of contingent payment models and the application of those models to insurance and other financial risks. A thorough knowledge of calculus, probability (as covered in Exam P), mathematical statistics (as covered in VEE Mathematical Statistics), interest theory (as covered in Exam FM), and the fundamentals of long-term actuarial mathematics (as covered in FAM) is assumed. This is a written-answer exam.

- Exam PA (Predictive Analytics): The syllabus for Exam PA provides candidates with the ability to employ selected analytic techniques to solve business problems and effectively communicate the solution. A thorough knowledge of probability (as covered in Exam P), mathematical statistics (as covered in VEE Mathematical Statistics), selected models and methods for analyzing data (as covered in Exam SRM) is assumed.

Validation by Education Experience (VEE)

- A candidate passes the VEE component of accreditation when the candidate shows that they have completed the coursework in specific subjects. After completing any two exams, candidates may apply to have the VEE credit earned applied to their records. The VEE requirements are completed by receiving a specified grade on VEE-approved college classes (The Society publishes on its website a list of approved courses offered by universities all over the world), online courses, or standard examinations. The VEE subjects are Economics, Accounting and Finance, and Mathematical Statistics.

E-learning courses

- Preactuarial foundations: This module introduces key skills in emotional intelligence (EQ) and adaptability (AQ), exposes the audience to the role of the actuary, and builds on technical expertise gained through prerequisite Exams P and FM. Self-awareness and communication are emphasized. The module culminates with an online end-of-module knowledge check, an interactive real-world scenario, and a take-home End-of-Module Assessment that asks students to provide written responses to open-ended questions.
- Actuarial science foundations: This module builds on the EQ and AQ skills introduced in the Preactuarial foundations module and leverages technical expertise gained through prerequisite Exams SRM and FAM (or equivalent). The module emphasizes the managing of ambiguity and introduces basic pricing techniques. The module culminates with an online end-of-module knowledge check, an interactive real-world scenario, and a take-home End-of-Module Assessment that asks students to provide written responses to open-ended questions.
- Fundamentals of actuarial practice: This online course encompasses real-world applications and examples to demonstrate actuarial principles and practices. The course presents practical techniques to assist in an actuary's day-to-day work.
- Advanced topics in predictive analytics (ATPA): ATPA provides candidates with the ability to employ selected data science and predictive analytics techniques to solve business problems and effectively communicate the solution. After learning this material from e-Learning modules, candidates are given a take-home assessment requiring analysis of a data set in the context of a business problem and submission of a report. A thorough knowledge of probability (as covered in Exam P), mathematical statistics (as covered in VEE Mathematical Statistics), and selected models and methods for analyzing data (as covered in Exams SRM and PA) is assumed.

Fellowship SOA examinations

According to the Society, Fellows of the Society of Actuaries must have demonstrated knowledge of the business environments within which financial decisions concerning retirement benefits, life insurance, annuities, health insurance, investments, finance, and enterprise risk management are made, including the application of advanced concepts and techniques for modeling and managing risk. In addition, Fellows must have in-depth knowledge of the application of appropriate concepts and techniques to a specific area of actuarial practice. Candidates for Fellowship must demonstrate that they possess this knowledge by passing a number of required courses in specific tracks. The Society describes these tracks as follows:

When an associate is ready to take the Fellowship requirements, they must select a specialty track and complete all requirements in that track. Mixing requirements from different tracks is not permitted. The Society offers six specialty tracks: Corporate Finance and ERM (CFE), General Insurance, Group and Health, Individual Life and Annuities, Quantitative Finance and Investment (QFI), and Retirement Benefits. The description of these six tracks gives you a very good idea of the mainstream activities of the members of the Society. Three exams and three e-learning modules are required for most specialty tracks, plus common capstone experience requirements – Decision Making and Communication (DMAC) and Fellowship Admissions Course (FAC), after completing all other exams and modules. General Insurance track requires four exams instead of three.

- DMAC is an e-learning module where the candidate acquires and uses knowledge that is distributed and facilitated electronically. There is also a project component. The focus of the DAMC Module is on written and oral communication skills and decision-making skills as they are applied to solving business problems.
- Fellowship Admissions Course (FAC)

As actuaries assume responsible positions within organizations, they need special skills and knowledge to carry out their new roles effectively. The FAC is designed to help actuaries deal effectively with the issues and situations they may confront as they progress in their organizations. It has several goals:

1. to increase awareness of professional ethical and malpractice issues and identify strategies to address them,
2. to encourage actuaries to approach problem-solving from varied directions and perspectives,
3. to provide coaching in oral communications, an enabling skill for actuaries.

As stated by the Society, the FAC can be taken any time after all other requirements have been completed. Successful completion of the course requires attendance and participation in all education sessions as well as successful completion of the oral presentation requirement.

SOA recommends FSA candidates take the exams and modules in a certain order, as presented here for each track.

Corporate Finance and ERM (CFE) track

After completing the CFE track, candidates will have a thorough understanding of the applications of economic theory to financial markets and modeling, advanced financial issues, and enterprise risk management. The track is designed to prepare candidates for employment in the fields of finance and enterprise risk management.

- ERM module: This is an e-learning module that provides candidates with an understanding of, and appreciation for, developing an ERM framework, identifying/defining operational risks, developing and analyzing economic capital models, and understanding various risk management approaches.
- ERM exam: The syllabus for this examination covers key topics in enterprise risk management and is a key component of the pathway for earning the Chartered Enterprise Risk Analyst (CERA) credential. The candidate will learn to understand, identitfy, analyze, measure, manage, and allocate risk using models and metrics.
- The Introduction to CFE module: This module provides candidates with an understanding and skills to evaluate a business through its strategy, financial statements, and the level of risk capital that it may need to hold. Candidates will also gain familiarity with authoritative resources that will facilitate future reading, reference, and practice, should that be needed for their work.
- Foundation of Corporate Finance and ERM exam: The syllabus covers advanced financial issues and enterprise risk management. The candidates are exposed to methods for risk identification, quantification and valuation, and techniques for risk management.
- Strategic decision-making exam: This is a 3.5-hour written-answer exam, offered in the spring and fall of each year. Topics covered on the exam include introduction to strategic management, strategic budgeting and value measures, decision modeling and optimization, and modeling complex systems. A case study is part of the examination.
- Advanced topics in CFE module: This module can include a broad range of concepts, which vary in their degree of complexity. This module covers three concepts which can be directly applied within a company's ERM framework. The candidate will learn about the factors that affect strategic thinking (external forces, environmental analysis), the organizational characteristics that influence strategic decision-making (strategy, structure, controls, leadership), and how senior management uses these to evaluate and benchmark progress toward strategic goals. Three applications of Extreme Value Theory (EVT) are covered to put the theory to work in a business context. The module leads a candidate through a strategic planning exercise while applying opportunity engineering (OE) concepts to help senior management make strategic decisions regarding a new product they would like to introduce to the market.

Individual Life and Annuities track

After completing the Individual Life and Annuities track, candidates will have a thorough understanding of individual life and annuity products. The aspects covered are financial reporting, valuation, product design, and pricing. The track is designed to prepare candidates for employment in the fields of financial reporting and product development.

- The Introduction to ILA FSA module: Actuaries play a critical role in life insurance companies to assess the probability of an event and its financial consequences. The Introduction to ILA module will explore topics including insurance products, financial reporting, reserving, distribution, and insurance administration.
- The Regulation and Taxation FSA Module is an e-learning module where candidates acquire an understanding of, and appreciation for, the regulatory environment that affects the life and annuity insurance industry overall and actuaries in particular.
- ERM module: Same as ERM module required for CFE track.
- Life Product Management exam: This is a written-answer examination offered in the spring and fall of each year. The syllabus for this exam covers the individual life insurance and annuity product design and development process. The course of reading exposes the candidate to the relationship between the product design and the selection of appropriate pricing assumptions as well as the profit measures used in pricing. This exam is also recognized by Canadian Institute of Actuaries.
- The Life Financial Management exam is a written-answer examination offered in the spring and fall of each year. Both a U.S. and Canadian version of the exam are administered, each of which has its own reading list. The syllabus for this exam exposes the candidate to the preparation and analysis of life insurance company financial statements including the underlying valuation principles supporting these statements. Further, the course of reading covers methods of reinsurance, risk mitigation, determining risk-based and economic capital, asset liability matching, and calculating embedded value for individual life insurance and annuity products.
- Life ALM and Modeling exam: Life ALM and Modeling Management is a two-hour written-answer exam that will be offered in the spring and fall of each year. The syllabus for the exam exposes candidates to the following topics: Asset and liability management and Enterprise risk management. The Enterprise Risk Management (ERM) Exam may be substituted for this exam.

Group and Health track

According to the SOA syllabus, after completing the Group and Health track, candidates will have a thorough understanding of group and health products.

The aspects covered are financial reporting, valuation, product design, and pricing. The track is designed to prepare candidates for employment in the fields of financial reporting, product development, and pricing.

- The Health Economics Module discusses the many stakeholders in the health care industry – patients, providers, government entities, employers, insurance companies – and the financial incentives that influence how they interact with the system.
- The Health Foundations Module is an e-learning module that discusses the health care system at a micro level. Through an understanding of terminology and coding, health care data and research can be understood, effectively assessed for quality, and effectively used in actuarial work.
- The Design and Pricing exam is a written-answer examination offered in the spring and fall of each year. The syllabus for this exam develops the candidate's knowledge of the design and pricing of group and health products. Specific topics include single employer group coverage, individual and multilife coverages, coverages offered through group vehicles, employer strategies for design and funding, government health plans, regulation, taxation, utilization and claims management, predictive modeling techniques, medical manual rates, credibility theory, and underwriting.
- The Finance and Valuation Exam is a written-answer examination offered in the spring and fall of each year. Both a U.S. and Canadian version of the exam will be administered, each of which has its own reading list. The syllabus for this exam develops the candidate's knowledge of the financial analysis of group and health products including the underlying valuation principles. Specific topics include reserving and actuarial appraisals, government programs, financial statement, regulations and taxation, and retiree benefits.
- The Group and Health Specialty exam is a two-hour written answer exam offered twice a year. Candidates who successfully complete the exam will be able to (1) understand, evaluate and mitigate risks assumed by health insurance organizations and plan sponsors; (2) demonstrate a full understanding of predictive modeling technique through applying them to underwriting, pricing, and care management situations; and (3) design and evaluate funding and accounting of a retiree medical and life plan. ERM exam can be substituted for this specialty exam.
- The Pricing, Reserving and Forecasting Module is an e-learning module where you acquire and use knowledge that is distributed and facilitated electronically. In this module, candidates will be exposed to practical techniques involved in managing the financial control cycle of a health care company – from trend determination to pricing and reserving to analysis of historical results to forecasting future experience. Candidates can take the ERM module as a substitute.

Quantitative Finance and Investment track

After completing this track, candidates will have a thorough understanding of portfolio management and the applications of economic theory to financial markets and modeling. The track is designed to prepare candidates for employment in the fields of investments and portfolio management.

- Enterprise risk management module as described previously.
- The Financial Modeling module is designed to provide candidates with extensive practice building and running the types of economic and capital market scenarios that may be required of an actuary. Candidates will gain familiarity with authoritative resources that will facilitate future reading, reference, and practice. The Financial Modeling module requires the candidate to have basic skills in using R and RStudio.
- The Scenario Modeling module is designed to provide candidates with extensive practice building and running the types of economic scenarios that may be required of an actuary. Candidates will also gain familiarity with authoritative resources that will facilitate future reading, reference, and practice. The Scenario Modeling module requires the candidate to have skills in using R and RStudio.
- Quantitative Finance and Investment Quantitative Finance Exam is a five-hour written-answer exam that is offered in the spring and fall of each year. Candidates are exposed to modern corporate finance theory and applications, advanced derivatives, and financial markets pricing and modeling. Specific topics include stochastic calculus, fixed income markets, interest rate models and hedging, equity option pricing and hedging, and their applications.
- The Portfolio Management exam is a five-hour written-answer examination offered in the spring and fall of each year. The syllabus for this exam covers the portfolio management process in detail. Candidates are exposed to fixed-income portfolio management, quantitative credit risk management and rating agency framework, equity and alternative investments, liquidity risk, investment policy and regulatory framework, asset liability management and asset allocation, performance measurement and attribution.
- Investment Risk Management is a two-hour written-answer exam that is offered in the spring and fall of each year. The syllabus of this exam covers governance in the context of investment operations and components of effective risk management. The Enterprise Risk Management (ERM) Exam may be substituted for this exam.

Retirement Benefits Track

After completing the Retirement Benefits track, candidates will have a thorough understanding of retirement plans. The aspects covered are plan design, funding, accounting, and valuation. This track prepares candidates for employment in the fields of defined benefit plans, defined contribution plans, and retiree health plans.

- The Social Insurance Module is an e-learning module. Employer-provided benefits should be designed to coordinate with social insurance programs. The Social Insurance programs vary widely in their coverage, cost, and benefits in different countries. It is critical that candidates be prepared to coordinate the employer-provided plans with the social insurance programs in their country. The Social Insurance Module goes into the details needed to practice as an actuary in the United States or Canada.
- The Design and Accounting exam is a five-hour written-answer examination offered in the spring and fall of each year. Both a U.S. and Canadian version of the exam are administered, each of which has its own reading list. The syllabus for this exam covers the basics of design and valuation for the full gamut of retirement plans found in the United States and Canada. Specific topics include benefit plan structure and features, regulation related to plan design, risk to plan sponsor and to plan participants, actuarial assumptions, plan funding and accounting standards, and professional standards of practice.
- The ERM Module as described previously.
- The Retirement Plan investment and Risk Management examination is a two-hour written-answer exam that is offered in the spring and fall of each year. Candidates will understand the issues facing retirement plan sponsors regarding investment of fund assets, recognize and appropriately reflect the role of plan investments in managing plan sponsor risk and make recommendations, and understand how to evaluate the stakeholders' financial goals and risk management with respect to their plan. The Enterprise Risk Management (ERM) Exam may be substituted for this exam.
- The Pension Projection module allows candidates to learn more about different modeling approaches available for pension projections, which model a pension plan over multiple periods. That is in contrast to a pension valuation which provides liabilities and assets at a single point in time. A pension projection typically includes modeling of pension assets together with pension liabilities. However, unfunded pay-as-you-go systems may not include any asset modeling. Candidates learn how to select assumptions, and potential ways to communicate the results of the projections.

General Insurance track

The SOA's General Insurance track is an NAIC (National Association of Insurance Commissioners) accepted actuarial designation, meeting NAIC standards of a qualified actuary in general insurance.

- The Introduction to General Insurance exam is a one-and-a-half-hour multiple-choice exam that is offered via computer-based testing (CBT). Through this exam, candidates will gain a greater understanding of the structure and functions of a general insurance company – underwriting roles, claim and reinsurance functions, risk control, liabilities, specialty coverages, and more.
- The Rate Making and Reserving exam is a five-hour written-answer exam that is offered in the spring and fall of each year. The candidate will understand the key considerations and concepts underlying general insurance actuarial work, including data preparation, projecting and trending ultimate claims, understanding financial reporting of liabilities, ratemaking, and catastrophe modeling.
- The Financial and Regulatory Environment exam: Financial and Regulatory Environment is a five-hour written-answer exam that will be offered in the spring and fall of each year. Candidates will understand the elements of financial reporting for general insurance companies, understand the analysis of a general insurer's financial health through prescribed formulas, ratios and other solvency regulation methods, and be able to apply the standards of practice regarding the responsibilities of the actuary as defined by regulators and the American Academy of Actuaries. Moreover, candidates will be able to describe the current and historical regulatory environment and be able to understand tort law and insurance law with respect to its impact on the general insurance industry.
- The Financial Economics module: Financial economics is the discipline underlying all financial services. No matter what the area of specialty or the role of an actuary, an understanding of financial economics is essential. Candidates will learn in this module that financial economics is the study of how individuals and institutions acquire, save, and invest money. These activities are fundamental to institutions and individuals.
- ERM module as described previously.
- The Applications of Statistical Techniques module introduces a set of advanced business analytic techniques to assist in key business decisions. While the emphasis is on understanding and implementing results, candidates are expected to always keep in mind the broader context of decision-making. After finishing this module, candidates are expected to master statistical techniques including generalized linear models, bootstrap, Bayesian method, and clustering method in various applications, and perform analysis using the statistical package R.
- The Advanced Topics in General Insurance exam is a two-hour written-answer examination that is offered in the spring and fall of each year. Topics covered on the exam include stochastic reserving, risk margins for unpaid claims, excess of loss coverages and retrospective rating, reinsurance pricing, and underwriting profit margin. The Enterprise Risk Management (ERM) Exam may be substituted for this exam.

Chartered Enterprise Risk Analyst Examinations

As is shown on the SOA website, another designation – the Chartered Enterprise Risk Analyst (CERA) – is available to help students and business professionals prepare for and seize opportunities in the evolving discipline of enterprise risk management (ERM) within broader financial services, insurance, and pension markets.

According to the SOA, the syllabus was developed to meet market needs and provides a rigorous treatment of critical ERM topics, including actuarial approaches to risk. The required examinations and modules are offered by SOA and consist of the following:

- Exam P (Probability)
- Exam FM (Financial Mathematics)
- Validation by Educational Experience (VEE) subjects: Economics, Accounting and Finance, and Mathematical Statistics.
- Preactuarial Foundations Module
- Exam FAM (Fundamental Actuarial Mathematics)
- Exam SRM (Statistics for Risk Modeling)
- Actuarial Science Foundations Module
- Fundamentals of Actuarial Practice
- Associateship Professionalism Course
- ERM Module
- Exam ERM (Enterprise Risk Management)

It is worth noting that candidates who have satisfied the CERA requirements can apply for Associateship of the SOA if they pass the Predictive Analytics exam and are approved by the SOA Board of Directors.

Continuing professional development

One of the important professional innovations introduced by the SOA is the Continuing Professional Development (CPD) requirement for its membership. According to the Society, all members must demonstrate on a regular basis that they are "SOA Compliant" by having met the SOA CPD requirement during the past two years, a requirement that began December 31, 2010.

According to the Society, members who are not compliant are those who failed to earn sufficient continuing education credits, elected not to comply with the requirement, or did not attest compliance.

The Society states that members do not lose their SOA designations (ASA, CERA, FSA) because they fail to comply with the SOA CPD requirement (although they can lose their designation if they have been found to violate the terms of noncompliance). On the other hand, all noncompliant members are required to tell users of their actuarial expertise of their failure to comply with the SOA CPD requirement.

The compliance component places considerable ongoing importance on one of the four cornerstones of the actuarial profession mentioned earlier: core knowledge, specialized knowledge, professional knowledge, and experiential knowledge.

2.5   The CAS Examinations

As in the case of SOA, the CAS syllabus has four corners: core knowledge, specialist knowledge, professional knowledge, and experiential knowledge. Accreditation proceeds at two levels: associates and fellows. Affiliate Membership recognizes that the Affiliate Member has been granted professional status as an actuary by another actuarial organization and practices in the property/casualty field, but does not meet the qualifications to become an Associate or Fellow of the CAS. Candidates become Associates if they meet the core and some specialist requirements, as well as some professional and experiential requirements. Let us take a closer look at the examinations involved.

Associateship requirements

Associateship Requirements include six examinations, credits in two VEE (validation by education experience) subjects, two online courses, and the CAS course on professionalism. The CAS recommends candidates accomplish the requirements in certain order, as shown.

- Exam 1 – Probability
  Currently Exam 1 required by the CAS is the same as Exam P by the SOA.

- Exam 2 – Financial Mathematics
  Currently the Financial Mathematics Exam of CAS is the same as Exam FM by the SOA.

- VEE Accounting and Finance: same as that of SOA
- VEE Economics: same as that of SOA
- CAS Data and Insurance Series Courses
  - Introduction to Data Analytics
  - Risk Management and Insurance Operations
  - Insurance Accounting, Coverage Analysis, Insurance Law, and Insurance Regulation

The online course Risk Management and Insurance Operation covers risk management, risk control, risk financing, enterprise risk management, insurance as risk management tool, insurance operation, insurance marketing and distribution, the underwriting function and underwriting cycle, underwriting property and liability insurance, claim function and adjusting claims, reinsurance, and actuarial data management. The CAS will grant a waiver of CAS Online Course 1 to those who have earned the Chartered Property Casualty Underwriter (CPCU) designation.

The online course Insurance Accounting, Coverage, Analysis, Insurance Law, and Insurance Regulation teaches candidates to understand introductory insurance accounting, policy analysis and common policy concepts, features of personal auto policy, homeowners' insurance, commercial property insurance, commercial general liability insurance, specialty coverages, and basics of life insurance, annuity, and health insurance. Insurance laws and regulations are also covered.

Each of the two online courses finishes with a 100-minute, 75-point multiple-choice examination. The online courses are available through the CAS Online Courses webpage.

- Exams I and II on Modern Actuarial Statistics (MAS-I, MAS-II)

Solid foundation in statistics is required of an actuary whose job function mainly pertains to property and casualty insurance (aka. general insurance). These required topics in statistics are covered by two four-hour examinations.

The MAS I exam covers probability models including stochastic process sand survival models, basic statistic estimation methods, hypothesis testing and properties of estimators, extended linear models, and time series with constant variance.

Topics covered in MAS II exam include credibility, linear mixed models, Bayesian analysis and Markov Chain Monte Carlo, and foundational statistical learning techniques such as k-nearest neighbors, decision tree, and principal components analysis.

- Exam 5 – Basic Techniques for Ratemaking and Estimating Claim Liabilities

This is a four-hour examination. The first section of the exam syllabus introduces the general principles of ratemaking as well as specific details regarding data requirements, calculations, key assumptions, and implementation-related issues. Candidates require a thorough understanding of basic ratemaking so that they will be able to analyze data, select appropriate techniques, and develop solutions to problems. The second section explores basic techniques that actuaries use to estimate unpaid claims for both insurance entities and also for noninsurance entities that retain risk. The American Academy of Actuaries' Standard of Practice related to the estimation of unpaid claims is also examined in the second section.

- Exam 6 – Regulation and Financial Reporting (Nation-Specific)

This is a 4-hour exam that focuses on nation-specific regulation and financial reporting standards applicable to property-casualty insurance companies.

- (United States)

The syllabus of the U.S. version of this course states that Section A of this examination covers insurance regulation with regards to property-casualty coverages, ratemaking, pricing, and solvency, and U.S. tort law as it affects the property-casualty business. Section B allows candidates to learn about the objectives, operation, and effectiveness of major insurance programs administered by government agency and insurance industry organizations, such as crop insurance, flood insurance, and residual markets of auto insurance and workers' compensation. Section C addresses financial reporting, solvency, and taxation issues. Candidates should have detailed knowledge of the contents, purposes, and recent changes in the NAIC Annual Statement and the Insurance Expense Exhibits. Knowledge of federal income tax treatment, including loss reserve discounting, is expected. Section D focuses on the professional responsibilities related to financial reporting of a property-casualty insurance company's appointed actuary according to the Property and Casualty Annual Statement Instructions issued by the National Association of Insurance Commissioners (NAIC). Section E presents the general concepts of reinsurance accounting to the candidate.

- (Canada)

The Canadian version of this four-hour examination assumes that candidates have completed the Online Course 2. This course contains the fundamental material for both the Regulations of Insurance and Canadian Insurance Law, as well as the Canadian financial reporting requirements.

Section A encompasses Canadian insurance legislation and regulations including their historical development. Judicial decisions affect insurance regulation and insurance benefits to the extent they interpret the law and thereby modify regulatory behavior. Section B focuses on the identification of major Canadian insurance programs administered by government agencies and insurance industry organizations. Section C addresses financial reporting and solvency issues. The intent is to address Canadian and global issues related to the reporting of financial results for property and casualty insurers. The core of the syllabus focuses on Canadian issues with an overview of relevant differences in other countries. Section D focuses on the professional responsibilities of the appointed actuary related to the reporting of financial results by property and casualty insurers in Canada.

- (Taipei)

The Actuarial Institute of Chinese Taipei (AICT) uses the Casualty Actuarial Society examinations for its property-casualty actuaries. The CAS Board of Directors approved specific AICT exams (i.e., current AICT Exam 6GA3 on Actuarial Standard of Practice and Accounting and Exam 6GB3 on Insurance Regulations and Discipline) as fulfilling the nation-specific requirement for CAS membership effective January 1, 2010.

- (International)

The International version of Exam 6 is developed to meet the needs of employers and future members in Asia, South America, and Europe. Candidates should understand the role of the insurance business as a supplier of a vital service. Because of the essential and highly technical nature of insurance, a system of regulatory controls has been established requiring insurers to demonstrate that they are providing fair and reliable services in accordance with the statutes and regulations of each jurisdiction. Candidates are expected to understand the concept and assessment of solvency, including ORSA (Own Risk and Solvency Assessment) and various international approaches to assessing solvency. Candidates are also expected to understand financial reporting under International Financial Reporting Standard 17 (IFRS 17). Candidates should understand the general concepts of reinsurance and become familiar with reinsurance accounting terminology and practice. Professional responsibilities of an actuary as defined by standards of practice, regulators, and insurance laws for financial reporting from an international viewpoint are also covered on this exam.

- **Professionalism course**

    The CAS also has a professionalism requirement. A candidate for the associateship must have successfully passed certain examinations, taken online courses, and earned VEE credits to be eligible for registration.

    Candidates who have completed all the requirements are eligible for Associateship of the CAS, namely ACAS. Three more exams are needed for the next level of CAS membership.

Fellowship examinations

- Exam 7 – Estimation of Policy Liabilities, Insurance Company Valuation, and Enterprise Risk Management

    Section A focuses on advanced techniques that the actuary may need to estimate reserves for unpaid claims, assuming that candidates are well versed in the basic Principles and Standards of Practice for unpaid claim estimation. This section addresses how actuarial concepts are adapted to evaluate liabilities arising in complex risk transfer agreements common in excess insurance and reinsurance contracts. Section B focuses on methods used to determine the theoretical value of equity securities and extending the methodology to value property and casualty insurance companies. Section C introduces the concepts and basic techniques of Enterprise Risk Management (ERM), a framework to integrate the entire landscape of risk that confronts a business.

- Exam 8 – Advanced Ratemaking

    Candidates for Exam 8 are expected to have already acquired considerable technical knowledge and practical experience in insurance ratemaking. Therefore, this examination will assume a working knowledge of basic ratemaking and will deal with advanced topics. Section A focuses on various aspects of classification rating, such as identifying and evaluating possible rate classes, estimating loss costs of rating classes, and building and using GLM models for classification ratemaking. Section B prepares candidates to design and manage excess, deductible, and individual risk rating systems. It covers pricing for layers of loss including excess and deductible business, experience rating in which prior individual risk experience is used to adjust rates prospectively, retrospective and loss-sensitive rating in which the insured will pay an amount (in premium or retained loss) that depends on the experience after the policy has been written. Section C covers catastrophe ratemaking and reinsurance ratemaking.

- Exam 9 – Financial Risk and Rate of Return

  Exam 9 focuses on a broad array of finance, investment, and financial risk management topics. This examination assumes a working knowledge of basic ratemaking, finance, probability and statistical modeling, liability and reserve risk, and insurance underwriting.

  Section A focuses on the portfolio theory and equilibrium in capital markets. Candidates are introduced to the manner in which investors might select a particular portfolio that best suits their individual preferences for risk and return. Candidates are presented to various equilibrium models, including the Capital Asset Pricing Model (CAPM) and Arbitrage Pricing Theory (APT). The concept of market efficiency is introduced to help candidates understand the factors that move market prices toward and away from the theoretical prices presented in these models. Section B exposes candidates to factors that influence the price sensitivity of fixed-income securities and presents various ways in which a portfolio manager might manage the interest rate and cash flow risk in a portfolio of these instruments. The same concepts are also applied to the interest rate risk associated with a firm's liabilities and the interest rate risk associated with a firm's total market value, inclusive of its franchise value. Section C addresses financial risks as well as risks related to the insurance industry from the financial economics perspective. Section D explores the relationship between insurance concepts (such as underwriting profits, premium-to-surplus ratios, and investment income) and financial concepts (such as interest rates, inflation rates, cost of capital, and risk premiums). The readings build on a background of finance as related to the insurance business, and deal with specific techniques used by actuaries to develop an appropriate profit loading in insurance prices.

General comments

If you take a closer look at the SOA and CAS exams that are required of associateship, you will discover that most of them deal with some aspect of mathematics and statistics in a business context. In addition to being able to carry out mathematical and statistical calculations, you must be able to understand the definitions and interrelationships of concepts from economics and finance. You will notice that many questions involve both definite, indefinite, and multiple integrals, as well as ordinary and partial derivatives. Hence, a good command of calculus is essential. Exponential and logarithmic functions are core functions, and so are geometric progressions. You will also discover that indispensable concepts from statistics are probabilities, distributions, random variables, expected values, mean, and variance and correlation. You will also notice that among the probability distributions, the Poisson distribution comes up most often for discrete variables. While the specific questions will obviously vary from year to year, it is clear from the nature of actuarial science that the mentioned ideas and techniques from mathematics, business, and statistics will always be part of the skills an actuary is required to possess.

2.6   Theory and Practice

Q: **What is the connection between the actuarial examinations and the knowledge and skills required in actuarial practice?**

ANS: The exams are good at introducing the material, but it really takes time in practice to gain a deep understanding of things. The exams help you learn the formula, but you learn how and when to use certain methods in practice.

ANS: Some like MLC/LTAM have materials which are applicable. As a pension actuary, the EA (Enrolled Actuary) exams are the most relevant.

ANS: Actuarial exams are a good preparation for actuarial practice, such as statistics, probability, financial mathematics, pricing, reserving, modeling, insurance regulation, and so on. These are all tested and useful in real work.

ANS: Exams provide the necessary fundamental knowledge to understand the models used in the workplace.

ANS: Actuarial examinations give you a baseline of knowledge at the beginning of your career. But toward the end of your actuarial exams (FSA), your job should give you a baseline for the exams.

ANS: There is a direct connection between the actuarial examinations and the knowledge and skills required in actuarial practice because the more actuary exams passed increase or provide the actuary with many tools to become an effective and efficient actuary.

ANS: The knowledge you learned from exams is somehow used in actuarial practice. Those exams are regularly updated due to the industry/regulation changes.

ANS: In the fellowship exams, actuaries acquire knowledge that is directly relevant to the industry they've selected. For instance, I found the life valuation exam helpful to understand the capital requirements that insurance companies are subject to under the Risk Based Capital framework.

ANS: Based on personal experience, there are quite some gaps between the actuarial examinations and the knowledge required in actual practice. The actuarial exams equip you with only high-level and simplified concepts, but in practice, there are always certain additional elements that complicate the standard situation, so you cannot directly apply the knowledge from the textbook. Additionally, in the examinations, all information is provided, but in reality, you may not even know where to find the updated information/regulation. In this case, shadowing a more experienced actuary and continued education can be helpful.

ANS: Some exams are more useful than others. There is a big connection between exams and the practice. Without the exams, one would not know what to do in practice.

ANS: The most important connection is the ability to learn new things on your own via studying and research.

ANS: Mainly – it's really in grain in the Casualty models, Increased Limit Factors, Loss Development Factors, reserving theories.

ANS: The exams provide a foundation for the real work. Very few actuaries will ever calculate a premium or reserve by hand on the job – this will almost always be done by a computer. However, the exam knowledge enables an actuary to understand what drives changes in the reserve – the computer does the leg work, and the actuary does the thinking about the final result. The best comparison I can think of is a mechanic – while a mechanic may have days full of oil changes and tire rotations, they still need to know how to diagnose a weird noise or spot when brake pads are wearing thin.

ANS: Outside of Exam 5, there really is no connection between the actuarial exams and the knowledge/skills required in my day-to-day work.

ANS: I would say the FSA modules are very relevant to healthcare. Basic reserving and pricing concepts are taught.

2.7   Difficulty of the Examinations

Q: Which SOA or CAS examinations did you find to be the most difficult and why? Illustrate your answer with examples.

ANS: CAS/SOA Exam 3 (IFM) and CAS MAS-I I found to be the most difficult. The amount of material was difficult to handle. The number of formulas and topics felt overwhelming at times.

ANS: MFE – lot of concepts that are not really used in my day-to-day work.

ANS: It's Exam 6 (CAS) for me since this one is the only one I failed until now. It has a lot of conceptual material which requires understanding deeply and memorizing. And it has a lot of practical knowledge like regulation, insurance laws, and practice of reinsurance. It includes a lot of stuff to learn.

ANS: Exam C – it had a massive syllabus with topics that did not intuitively relate to each other.

ANS: I found finance and valuation to be the most difficult. I am currently working on it, and it has given me the most trouble. I think the FSA exams concentrate a lot on lists and laws/regulations in addition to calculations.

ANS: I am still in the early stage of my actuary career where I still have not passed any SOA or CAS exam yet, so subjectively, I believe that all SOA or CAS exams are difficult because with my experience so far, it is difficult to pass SOA or CAS exam if you do not have a strong conceptual and analytic understanding of the particular exam there is no way you would be successful. This is because the exams do not only test your conceptual understanding but also the speed of recalling and solving the questions on the exam.

ANS: CAS upper-level exams are hard for me, especially when they switch to CBT (computer-based test) and there is no personal exam report available after each exam. When you fail, you don't know where you should improve yourself. Another reason is that there is not enough time for all the questions. You need to skip some if necessary, but not too many or else you will fail.

ANS: Predictive analytics. Have to learn and memorize the code and practice a lot.

ANS: ILALFV – which is the valuation exam under life and annuities. Because I started out in the pensions industry and many of the life insurance valuation concepts were new to me.

ANS: To me, SOA Life Product Management Exam and Life Financial Management Exam are the most difficult. Many concepts, such as insurance tax, regulations, value creations, advanced reinsurance, asset/ portfolio management, etc. are new and in-depth but still very much related to an actuary's responsibilities.

ANS: CAS Exam 6U – Regulation and Financial Reporting (US). This was most difficult because the concepts and learning material were less about mathematical theory, and more about regulation and financial reporting. It tested the exam taker's ability to recall information as it pertained to the rules and regulations associated with Financial Reporting. For example, the user had to memorize the history of State versus Federal regulation of Insurance within the United States.

ANS: So far I have found Exam MAS-I to be the most difficult given all the statistical concepts.

ANS: CAS Exam 6 was the most difficult, as it requires a lot of memorization of insurance accounting concepts and also insurance regulation history.

ANS: All of them :) I'm not the best exam taker, and the exams are too lengthy without a focus, and too many topics on any singular exam, while trying to study them all, but not all of them are tested. But if I have to select one exam to be most difficult, it would be Exam 6 due to the unfamiliar concepts and less math in that exam.

ANS: The SOA's Final Assessment as part of the Fundamentals of Actuarial Practice modules was the most difficult exam in my opinion. The assessment itself is quite difficult, and the grading is highly subjective. It was also impossible to get meaningful feedback on what you did wrong, and because of the open-endedness of the assessment it was impossible to just "study harder" for it. If you failed any other exam, you could tell which sections you needed to study harder. The FAP was not like that at all.

ANS: CAS Exam 6 has been the most difficult for me so far. This is due to the fact that there is a lot of material to it, and a lot of the exam questions include exceptions to certain rules that you have to know more than the other exams I've sat for.

ANS: I found exam MFE to be the most difficult because for me it was the first time I've ever been exposed to complex financial math.

ANS: Exam 6.

2.8   Study Material and Resources

The ACTEX Learning/Mad River Books (www.actexmadriver.com) is a publisher and distributor of textbooks, study manuals, and other learning tools, specializing in actuarial education. Actuarial Bookstore sells comprehensive sets of actuarial examination tools, for both the SOA and CAS examinations. Details can be found on the bookstore website at www.actuarialbookstore.com/.

In addition, the ASM study materials are available at www.studymanuals. com. They are also marketed by ACTEX.

Online tutoring is gaining popularity among actuarial students, as can be seen from the survey results:

- Coaching actuaries: www.coachingactuaries.com
- The Infinite Actuary: https://www.theinfiniteactuary.com/

2.9   Ways to Pass Examinations

Q : **What were your study tricks and study processes that helped you pass your actuarial examinations? Illustrate your answer with examples related to the SOA and CAS examinations.**

ANS: Finding the right study program helped me a great deal. I am a visual learner, so when I started using The Infinite Actuary I began understanding the material better which led to passing more exams. I also realized that studying too much is not beneficial. I used to burn myself out by studying too far out of my exam. Studying less has actually helped me pass more exams.

ANS: Study every day for a few hours, knowing that failing an exam happens.

ANS: For the exam P/FM/IFM/MAS-I/Exam 5 (CAS), my trick is to do a lot of practicing. I used TIA for MAS-I and Exam 5 and I did all the practicing questions in every chapter. It really helped me nail the exams. For MAS-II, I read the source material multiple times, which was highly recommended. My study process is that I would set up goals and everyday tasks and stick to them.

ANS: Reviewing problems that I had trouble with to improve on my weaknesses.

ANS: I think doing practice problems is by far the most important study tool. The exam problems are very similar to practice problems. I do not believe the difficulty is the issue for many students, I believe they struggle with the speed to complete the problems.

ANS: I am studying for SOA Exams P and FM, so my strategies include solving 10 quizzes a day using coaching actuaries, TIA study manuals, and ACTEX FM and P.

ANS: Practice a lot with timer. You don't have time to think when you deal with upper-level exams, and you need to finish each question in an average of six minutes. Put a timer on when you practice. If you don't have any ideas when you see the practice questions, that means you aren't ready and you need to study more.

ANS: Do not stay with the concept if you feel it is hard to understand, skip and come back later.

ANS: Studying at a library rather than at home helped ensure that distractions were not within reach. I tended to start with study manuals, and then supplement with self-paced seminars because I'm a visual learner and having and instructor whiteboard concepts is helpful to me.

ANS: My regular study process for FSA exams is to go over the detailed study manuals to grasp the high-level ideas and then the summarized PowerPoint slides for more details. I'd go over the study manuals/PowerPoint slides multiple times and around 30 days prior to the exam, I started to memorize the TIA flash cards. Each FSA exam preparation is around 2.5 months.

ANS: Creating a schedule/timeline always helps. Studying before work also helps in having the mental capacity to study. Having a study buddy, coffee, and snacks will make studying easier.

ANS: It is crucial to practice sample questions, and also try to make up sample questions for newer material that does not have good sample question coverage.

ANS: Doing a lot of questions is really the only way to drill the concepts into your head.

ANS: I made sure to take all the study hours allowed by my company, plus many more additional hours of my own time. In the weeks leading up to an exam, I would often work a 6- to 7-hour day, then study for 2 to 3 hours a night on the weekdays. On the weekends, I would study for 3 to 4 hours one day, and then almost always taking the other day to rest. If I came across a topic that I particularly struggled with, I'd first try and re-read that chapter from the very beginning. If rereading didn't help, then I'd backtrack and see if there was another topic that I was perhaps not so confident about that was affecting my understanding of this one. And if that still didn't work, then I'd look for alternate explanations of the topic – either in the source material, or even online on sites like Wikipedia and YouTube.

ANS: My preferred study trick is to take as many practice questions as possible, since I learn better through making mistakes than by reading through material.

ANS: For me the most helpful way to pass exams is going to study classes and asking my tutor to explain the basic concepts that will help answer any actuarial question. By just studying old questions, you close yourself from potential new ways in which the questions could be asked.

ANS: Read text or watch videos, understand the concepts, do problems, practice, practice, and practice.

2.10   Study Aids

Actuarial examinations are difficult to pass. The average pass rate is usually below 40%. This means, of course, that the average failure rate is often more than 60%. In other words, many bright and dedicated students who are used to getting high grades in college have to adjust to the fact that they may actually fail an examination. To help increase the students' chances of success in actuarial examinations, an extensive commercial support system has been developed. Various companies market different types of study tools, organize seminars and special courses, and provide other help for a fee.

Q: **Which study aids, such as ACTEX, have you used and which would you recommend? Illustrate your answer with examples related to the SOA and CAS examinations.**

ANS: The Infinite Actuary from CAS Exams MAS-I, MAS-II, 5, 6, and 8. This program helped me as a visual learner.

ANS: ACTEX and ASM manuals are helpful as are coaching actuaries for their practice exams and the infinite actuary. However, those are very expensive.

ANS: Exam P (ASM), FM (ASM), IFM (ACTEX), MAS-I (TIA), MAS-II (TIA), Exam 5 (TIA), Exam 6 (Battle Acts).

ANS: TIA for FSA exams, I hear adapt is great for lower level exams.

ANS: I believe coaching actuaries is the best for ASA exams. The practice problem quizzes are immensely helpful. I think TIA is best for FSA because they provide notecards for memorizing.

ANS: I am currently using Coaching Actuaries (CA), The Infinite Actuary (TIA) study manuals, and ACTEX FM and P.

ANS: For lower-level exams, I recommend coachingactuary.com website and ASM. Use coachingactuary.com for practice and ASM for reading material. For upper-level exams, I don't quite like TIA, but I don't have other good alternatives.

ANS: Coaching Actuaries.

ANS: ASM for study manuals, as they tended to be more in depth. The Infinite Actuary has very useful self-paced seminars.

ANS: To study for FSA exams, I used the TIA study manuals and rarely referred to the original study notes. Study manuals are more organized and structured in a way that is easier to follow and comprehend. However, the original study notes are more helpful in real work with all the necessary details. TIA also has video classes for each topic, which I'd only refer to for concepts that are difficult to understand on my own. What I found the most helpful is the TIA mobile app, especially the flashcard feature. It also helps to set the study schedule/discuss questions in the class forum/study video lessons so that you can stay on track for the exam.

ANS: Coaching Actuaries and Infinite Actuaries. Coaching actuaries is only available for the earlier CAS exams.

ANS: I think it is usually best to select study aids with the highest volume and quality of sample questions.

ANS: The infinite actuary had been my favorite.

ANS: I often used the ASM study manuals as my main studying resource, accompanied by Coaching Actuaries' practice quizzes and tests. I would recommend both of these as Dr. Weishaus does a very good job of explaining the topics, and Coaching Actuaries' platform is very intuitive and helpful. I'd also occasionally use the internet and sites like Wikipedia and YouTube to get explanations for particularly tricky topics.

ANS: Coaching Actuaries for all of my preliminary SOA exams. Infinite Actuary for FSA exams.

ANS: I have used ACTEX, Coaching Actuaries, and The Infinite Actuary. I would recommend Coaching Actuaries since you can create custom quizzes of various difficulty of whatever length you want. Also, the solutions to the questions are explained quite well.

ANS: I would recommend an actuarial tutor if you want someone to tutor you. And I would recommend ACTEX/Infinity Actuaries and Coaching actuaries.

ANS: Coaching actuaries, TIA. They both are good.

2.11 Mathematical Foundations of Actuarial Science – Probability and Calculus

Ideas and techniques

The world around us is full of uncertainty, some of which results in financial losses to us. We call such uncertainty risk. Insurance is a financial instrument to transfer certain risk to the insurance company. Actuaries are experts in pricing insurance products. The job essentially put a price tag on uncertainty. The basic tool is probability theory and applications. Probability theory was born to deal with a particular uncertainty. It is said that a professional gambler named Chevalier de Mere made a great deal of money by betting people that by rolling a die four times, he could get at least one six. He was so successful at it that he soon had trouble finding people willing to play his game. So he changed the rules. He started to bet that he could get at least two sixes by rolling a die twenty-four times. Unfortunately for him, he systematically lost. He contacted Pascal (1623–1662) to help explain his losses. Pascal began to correspond with Fermat (1601–1665) to analyze the problem and it is said that thus probability theory was born.

Not long after the birth of probability theory, calculus, a tool greatly useful for solving probability questions that involve continuous random variables, was invented and developed by Newton (1642–1727) and Leibniz (1646–1716).

Exam P of SOA (Exam 1 by CAS) helps candidates develop knowledge of the fundamental probability tools for quantitatively assessing risk, with emphasis on the application of these tools to problems encountered in actuarial science. A thorough command of the supporting calculus is needed. Additionally, a very basic knowledge of insurance and risk management is assumed.

Examination topics

The examination consists of 30 multiple-choice questions. They deal with the following topics from the SOA and CAS syllabus:

Part I: Basic probability concepts
The candidate will be able to:

(a) Define set functions, Venn diagrams, sample space, and events. Define probability as a set function on a collection of events and state the basic axioms of probability.
(b) Calculate probabilities using addition and multiplication rules.
(c) Define independence and calculate probabilities of independent events.
(d) Calculate probabilities of mutually exclusive events.
(e) Define and calculate conditional probabilities.
(f) Calculate probabilities using combinatorics, such as combinations and permutations.
(g) State Bayes Theorem and the law of total probability and use them to calculate conditional probabilities

Part II: Univariate random variables

The candidate will understand key concepts concerning discrete and continuous univariate random variables (including binomial, negative binomial, geometric, hypergeometric, Poisson, uniform, exponential, Gamma, normal, and mixed) and their applications, specifically the candidate will be able to:

(a) Explain and apply the concepts of random variables, probability and probability density functions, and cumulative distribution functions.
(b) Calculate conditional probabilities.
(c) Explain and calculate expected value, mode, median, percentile, and higher moments.
(d) Explain and calculate variance, standard deviation, and coefficient of variation.
(e) Define probability generating functions and moment generating functions and use them to calculate probabilities and moments.
(f) Determine the sum of independent random variables (Poisson and normal).
(g) Apply transformations.

Part III: Multivariate random variables

The candidate will understand key concepts concerning multivariate random variables (including the bivariate normal) and their applications.

The candidate will be able to:

(a) Explain and perform calculations concerning joint probability functions, probability density functions, and cumulative distribution functions.
(b) Determine conditional and marginal probability functions, probability density functions, and cumulative distribution functions.
(c) Calculate moments for joint, conditional, and marginal random variables.
(d) Explain and apply joint moment generating functions.
(e) Calculate variance, standard deviation for conditional and marginal probability distributions.
(f) Calculate joint moments, such as the covariance and the correlation coefficient.
(g) Determine the distribution of a transformation of jointly distributed random variables.
(h) Determine the distribution of order statistics from a set of independent random variables.
(i) Calculate probabilities and moments for linear combinations of independent random variables.
(j) State and apply the Central Limit Theorem.

Exam P sample questions and answers

Here are some examples of how these ideas and techniques can be tested.

Q1 Binomial distribution, bonus scheme.
Q2 Medical insurance, claims, independence, probabilities of union events, and intersection events.

Q3 Health plan, probability of mutually exclusive events, and complement events.

Q4 Probabilities of union events, mutually exclusive events, and complement events.

Q5 Total probability formula, probability of independent events.

Q6 Auto insurance, age as risk classification factor, Bayes theorem.

Q7 Game designed by cases, probability using combination.

Q8 Auto insurance, modeling of claim number, geometric sequence.

Q9 Distribution density function, conditional probability.

Q10 Time-to-failure, exponential distribution, conditional probability.

Q11 Health care plan design by case, cumulative distribution function, conditional probability.

Q12 Lifetime, exponential distribution, function defined by case, expected value.

Q13 Insurance, deductible, loss of uniform distribution, expected value.

Q14 Poisson distribution, function defined by case, mean, variance, and standard deviation.

Q15 Profit, normal distribution, standard deviation, ratio of probabilities.

Q16 Loss function defined by case, insurance with deductible, expected value of a not covered loss.

Q17 Function defined by case, binomial distribution, expected revenue.

Q18 Moment generating function, standard deviation.

Q19 Density function defined by case, variance of function of random variable.

Q20 Automobile loss, insurance with deductible, normal approximation of sum of variables.

Q21 Two independent variables, uniform distribution, graph of functions, areas.

Q22 Waiting time, independent exponential distributions, probability of joint events.

Q23 Employee benefit plan, joint distribution density function defined by case, expected value of marginal distribution.

Q24 Time-to-failure, joint distribution density function defined by case, expected value of marginal distribution.

Q25 Joint distribution density function defined by case, expected value of a function of bivariate.

Q26 Automobile liability and collision insurance, joint distribution density function defined by case, variance of marginal distribution.

Q27 Discrete joint distribution, number of tornadoes, conditional probability.

Q28 Employee benefit plan, joint distribution of taking-up rate, conditional variance.

Q29 Lifetime, composite function as cumulative distribution function, density function, chain rule of differentiation.

Q30 Vison care insurance, number of claims, Poisson distribution, normal approximation, sum of random variables.

Question 1 *A company establishes a fund of 120 from which it wants to pay an amount, C, to any of its 20 employees who achieve a high-performance level during the coming year. Each employee has a 2% chance of achieving a high-performance level during the coming year, independent of any other employee.*

Determine the maximum value of C for which the probability is less than 1% that the fund will be inadequate to cover all payments for high performance.

Answer Let X denote the number of employees who achieve a high-performance level. Then X follows a binomial distribution with parameters $n = 20$ and $p = 0.02$. The total payments f\or high performance is CX. We need to determine the maximum value of X such that $P(CX \geq 120) < 0.01$, or equivalently $P(CX < 120) \geq 0.99$, or $P\left(X < \dfrac{120}{C}\right) \geq 0.99$.

Given that X follows binomial distribution with parameters $n = 20$ and $p = 0.02$, we have

$$P(X = k) = \binom{20}{k}(0.02)^k (0.98)^{20-k} \text{ for } k = 0, 1, 2, \ldots, 20$$

The first three probabilities (at $k = 0,1,2$) are 0.668, 0.272 and 0.053, that makes $P(X \leq 2) = 0.993$. So the smallest X value that satisfies $P(X \leq k) > 0.99$ is 2. Thus the maximum value of C is obtained by setting $\dfrac{120}{C} = 2$, therefore $C = 60$.

Question 2 *An insurance company pays hospital claims. The number of claims that include emergency room or operating room charges is 85% of the total number of claims. The number of claims that do not include emergency room charges is 25% of the total number of claims. The occurrence of emergency room charges is independent of the occurrence of operating room charges on hospital claims.*

Calculate the probability that a claim submitted to the insurance company includes operating room charges.

Answer Let E = occurrence of emergency room charges, and O = occurrence of operating room charges. We know immediately that $P(E \cup O) = 0.85$ and $P(E^c) = 0.25$.

Using $P(E \cup O) = P(E) + P(O) - P(E \cap O)$, independence of E and O, and $P(E) = 1 - P(E^c) = 0.75$,

We have $0.85 = 0.75 + P(O) - (0.75)*P(O)$, and it follows that
$$P(O) = \frac{0.85 - 0.75}{1 - 0.75} = 0.40.$$

Question 3 *An insurer offers a health plan to the employees of a large company. As part of the plan, the individual employees may choose exactly two of the supplementary coverages A, B, and C, or they may choose no supplementary coverage. The proportions of the company's employees that choose coverages A, B, and C are 1/4, 1/3, and 5/12, respectively.*

Determine the probability that a randomly chosen employee will choose no supplementary coverage.

Answer Let x be the probability of employees who choose A and B, but not C, y the probability of employees who choose A and C, but not B, and z the probability of employees who choose B and C, but not A. It is given that
$$x + y = \frac{1}{4}, x + z = \frac{1}{3}, y + z = \frac{5}{12}.$$

We need to find $w = 1-(x + y + z)$.

Adding the three equations above, we have $2(x + y + z) = \frac{1}{4} + \frac{1}{3} + \frac{5}{12} = 1.$

Therefore $w = 1 - (x + y + z) = 1 - \frac{1}{2} = \frac{1}{2}.$

Question 4 *A survey of 100 TV watchers revealed that over the last year:*
(i) 34 *watched CBS.*
(ii) 15 *watched NBC.*
(iii) 10 *watched ABC.*
(iv) 7 *watched CBS and NBC.*
(v) 6 *watched CBS and ABC.*
(vi) 5 *watched NBC and ABC.*
(vii) 4 *watched CBS, NBC, and ABC.*
(viii) 18 *watched HGTV and of these, none watched CBS, NBC, or ABC.*

Calculate how many of the 100 TV watchers did not watch any of the four channels (CBS, NBC, ABC, or HGTV).

Answer Let

C = the number of TV watchers who watched CBS over the last year
N = the number of TV watchers who watched NBC over the last year
A = the number of TV watchers who watched ABC over the last year
H = the number of TV watchers who watched HGTV over the last year

Then the set $C \cup N \cup A$ represents the number of those who watched at least one from CBS, NBC, and ABC. Its number is $34 + 15 + 10 - 7 - 6 - 5 + 4 = 45$.

Because $C \cup N \cup A$ and H are mutually exclusive, the number of TV watchers who watched one from CBS, NBC, ABC, and HGTV is $45 + 18 = 63$, meaning the number of TV watchers who did not watch any of the four named channels is $100 - 63 = 37$.

Question 5 *An urn contains 10 balls: 4 red and 6 blue. A second urn contains 16 red balls and an unknown number of blue balls. A single ball is drawn from each urn. The probability that both balls are the same is 0.44.*
   Calculate the number of blue balls in the second urn.

Answer

$P(\text{both balls are the same color}) = P(\text{both balls are red}) + P(\text{both balls are blue})$
$$= P(\text{ball from urn 1 is red})P(\text{ball from urn 2 is red})$$
$$+ P(\text{ball from urn 1 is blue})P(\text{ball from urn 2 is blue})$$

Let x be the number of blue balls in the second urn.

We have $0.44 = \dfrac{4}{10}\dfrac{16}{16+x} + \dfrac{6}{10}\dfrac{x}{16+x}$. The solution is $x = 4$.

Question 6 *An auto insurance company insures drivers of all ages. An actuary compiled the following statistics on the company's insured drivers (see table).*

Age of driver	Probability of accident	Portion of company's insured drivers
16–20	0.06	0.08
21–30	0.03	0.15
31–65	0.02	0.49
66–99	0.04	0.28

A randomly selected driver that the company insures has an accident. Calculate the probability that the driver was age 16 to 20.

Answer This is an application of Bayesian theorem.

Let A = event of an accident

B_1 = event the driver's age is in the range $16 - 20$

B_2 = event the driver's age is in the range $21 - 30$

B_3 = event the driver'sage is in the range 31 – 65

B_4 = event the driver'sage is in the range 66 – 99

Then

$$P(B_1|A) = \frac{P(A|B_1)P(B_1)}{P(A|B_1)P(B_1) + P(A|B_2)P(B_2) + P(A|B_3)P(B_3) + P(A|B_4)P(B_4)}$$

$$= \frac{(0.06)(0.08)}{(0.06)(0.08) + (0.03)(0.15) + (0.02)(0.49) + (0.04)(0.28)} = 0.1584.$$

Question 7 *In a casino game, a gambler selects four different numbers from the first twelve positive integers. The casino then randomly draws nine numbers without replacement from the first twelve integers. The gambler wins the jackpot if the casino draws all four of the gambler's selected numbers.*

Calculate the probability that the gambler wins the jackpot.

Answer This question is equivalent to "What is the probability that 9 different chips randomly drawn from a box containing 4 red chips and 8 blue chips will contain the 4 red chips?" The hypergeometric probability is

$$\frac{\binom{4}{4}\binom{8}{5}}{\binom{12}{9}} = \frac{1*56}{220} = 0.2545.$$

Question 8 *In modeling the number of claims filed by an individual under an automobile policy during a three-year period, and actuary makes the simplifying assumption that for all integers $n \geq 0, p_{n+1} = 0.2 p_n$ where p_n represents the probability that the policyholder files n claims during the period.*

Under this assumption, what is the probability that a policyholder files more than one claim during the period?

Answer

For any $n \geq 0$, $p_n = (0.2)p_{n-1} = (0.2)^2 p_{n-2} = \ldots = (0.2)^n p_0$.

$$1 = \sum_{n=0}^{\infty} p_n = \sum_{n=0}^{\infty} (0.2)^n p_0 = \frac{p_0}{1-0.2}, thus \ p_0 = 0.8.$$

Finally $P(\text{more than one claims}) = 1 - P(0 \text{ claims}) - P(\text{one claim})$
$$= 1 - (0.8) - (0.2)(0.8) = 0.04.$$

Question 9 *An insurance company insures a large number of homes. The insured value, X, of a randomly selected home is assumed to follow a distribution with density function*

$$f(x) = \begin{cases} 3x^{-4} & \text{for } x > 1 \\ 0 & \text{otherwise} \end{cases}$$

Given that a randomly selected home is insured for at least 1.5, what is the probability that it is insured for less than 2?

Answer The cumulative distribution function is

$$F(x) = P(X \le x) = \int_1^x 3t^{-4}dt = -t^{-3}\big|_1^x = 1 - x^{-3}, \text{ when } x > 1$$

Thus $P(X < 2 | X \ge 1.5) = \dfrac{P(1.5 \le X < 2)}{P(X \ge 1.5)} = \dfrac{F(2) - F(1.5)}{1 - F(1.5)}$

$$= \frac{\left(1 - 2^{-3}\right) - \left(1 - 1.5^{-3}\right)}{1 - \left(1 - 1.5^{-3}\right)} = \frac{1.5^{-3} - 2^{-3}}{1.5^{-3}}$$

$$= 1 - \left(\frac{4}{3}\right)^{-3} = \frac{37}{64} = 0.578.$$

Question 10 *A car is new at the beginning of a calendar year. The time, in years, before the car experiences its first failure is exponentially distributed with mean 2.*

Calculate the probability that the car experiences its first failure in the last quarter of some calendar year.

Answer Using the law of total probability, the sought-after probability is

$$\sum_{k=0}^{\infty} P(k + 0.75 < X \le k + 1 | k < X \le k + 1) * P(k < X \le k + 1)$$

The first term in the product is

$$P(k + 0.75 < X \le k + 1 | k < X \le k + 1) = \frac{P(k + 0.75 < X \le k + 1)}{P(k < X \le k + 1)}$$

$$= \frac{F(k + 1) - F(k + 0.75)}{F(k + 1) - F(k)}$$

$$= \frac{\left(1 - e^{-(k+1)/2}\right) - \left(1 - e^{-(k+0.75)/2}\right)}{\left(1 - e^{-(k+1)/2}\right) - \left(1 - e^{-k/2}\right)}$$

$$= \frac{e^{-0.375} - e^{-0.5}}{1 - e^{-0.5}} = 0.205.$$

Since this term is constant, regardless of the value of k, the total probability is equal to

$$\left(0.205\right)\sum_{k=0}^{\infty}P\left(k < X \le k+1\right) = \left(0.205\right)P\left(X > 0\right) = \left(0.205\right).$$

Question 11 *The cumulative distribution function for health care costs experienced by a policyholder is*

$$F\left(x\right) = \begin{cases} 1 - e^{-\frac{x}{100}}, & \text{for } x > 0 \\ 0, & \text{otherwise} \end{cases}$$

The policyholder has a deductible of 20. An insurer reimburses the policy-holder for 100% of the health care costs between 20 and 120. Health care costs above 120 are reimbursed at 50%.

Let G be the cumulative distribution function of reimbursements, given that the reimbursement is positive.

Calculate G (115).

Answer Let Y be the reimbursement amount. Then

$$Y = \begin{cases} 0, & X \le 20 \\ X - 20, & 20 < X \le 120 \\ 100 + 0.5\left(X - 120\right), & X > 120 \end{cases}$$

and $G\left(115\right) = P\left(Y \le 115 | X > 20\right)$.

For Y to be 115, X must be 150. Therefore

$$G\left(115\right) = P\left(X \le 150 | X > 20\right) = \frac{P\left(X \le 150\right) - P\left(X \le 20\right)}{P\left(X > 20\right)}$$

$$= \frac{\left(1 - e^{-\frac{150}{100}}\right) - \left(1 - e^{-\frac{20}{100}}\right)}{1 - \left(1 - e^{-\frac{20}{100}}\right)} = \frac{-e^{-1.5} + e^{-0.2}}{e^{-0.2}} = 1 - e^{-1.3} = 0.727.$$

Question 12 *The lifetime of a printer costing 200 is exponentially distributed with mean 2 years. The manufacturer agrees to pay a full refund to a buyer if the printer fails during the first year following its purchase, and a one-half refund if it fails during the second year.*

If the manufacturer sells 100 printers, how much should it expect to pay in refunds?

Answer Let T be the printer's lifetime, time from purchase to failure. The distribution is $F(t) = 1 - e^{-\frac{t}{2}}$. The probability of failure in the first year is $P(T \le 1) = F(1) = 0.3935$, and the probability of failure in the second year is $P(1 < T \le 2) = F(2) - F(1) = 0.6321 - 0.3935 = 0.2386$.

Of 100 printers, the expected number of failures in the first year, and in the second year is 39.35, and 23.86, respectively. Thus the total expected refunds is

$$200(39.95) + 100(23.86) = 10,256.$$

Question 13 *An insurance policy is written to cover a loss, X, where X has a uniform distribution on [0,1000].*
At what level must a deductible be set in order for the expected payment to be 25% of what it would be with no deductible?

Answer With no deductible, the expected payment is 500, so the desired expected payment with deductible is 125. Let d be the deductible.

$$125 = \int_d^{1000} (x - d)(0.001)\,dx = (0.001)\frac{(x-d)^2}{2}\Big|_d^{1000} = 0.0005\left[(1000 - d)^2 - 0^2\right]$$

Solve for d. We have $d = 500$.

Question 14 *A baseball team has scheduled its opening game for April 1. If it rains on April 1, the game is postponed and will be played on the next day that it does not rain. The team purchases insurance against rain. The policy will pay 1000 for each day, up to 2 days, that the opening game is postponed.*
The insurance company determines that the number of consecutive days of rain beginning on April 1 is a Poisson random variable with mean 0.6.
What is the standard deviation of the amount the insurance company will have to pay?

Answer Let N be the number of consecutive days of rain beginning on April 1. The expected amount paid is

$$1000P(N = 1) + 2000P(N > 1) = 1000\frac{(0.6)e^{-0.6}}{1!}$$
$$+ 2000\left(1 - e^{-0.6} - \frac{(0.6)e^{-0.6}}{1!}\right) = 573.$$

The second moment of the amount paid is

$$1000^2 P(N = 1) + 2000^2 P(N > 1)$$

$$= 1000^2 \frac{(0.6)e^{-0.6}}{1!} + 2000^2 \left(1 - e^{-0.6} - \frac{(0.6)e^{-0.6}}{1!} \right) = 816,893$$

The standard deviation is $\left[816,893 - 573^2 \right]^{0.5} = 488,564^{0.5} = 699.$

Question 15 *Insurance companies A and B each earn an annual profit that is normally distributed with the same positive mean. The standard deviation of company A's annual profit is one-half of its mean.*

In a given year, the probability that company B has a loss (negative profit) is 0.9 times the probability that company A has a loss.

Calculate the ratio of the standard deviation of company B's annual profit to the standard deviation of company A's annual profit.

Answer Let X and Y represent the annual profit for companies A and B, respectively, and m be the common positive mean, then the standard deviation of X will be $\dfrac{m}{2} = 0.5m.$ Let s be the standard deviation of Y. Use Z to denote the standard normal distribution.

Using standardization of normal distribution, company A has a loss at the probability

$$P(X < 0) = P\left(\frac{X - m}{0.5m} < \frac{0 - m}{0.5m} \right) = P(Z < -2) = 0.0228.$$

And company B has a loss at the probability:

$$P(Y < 0) = P\left(\frac{Y - m}{s} < \frac{0 - m}{s} \right) = P(Z < -m/s).$$

Since $P(Y < 0) = 0.9\, P(X < 0),$ we have $P(Z < -m/s) = 0.9(0.0228) = 0.02052.$

From the table for standard normal distribution (given on exam), $-2.04 = -\dfrac{m}{s},$ or $s = \dfrac{m}{2.04}.$

The ratio of two standard deviations is $\dfrac{\frac{m}{2.04}}{0.5m} = 0.98.$

Question 16 *A manufacturer's annual losses follow a distribution with density function*

$$f(x) = \begin{cases} \dfrac{2.5(0.6)^{2.5}}{x^{3.5}}, & \text{for } x > 0.6 \\ 0, & \text{otherwise} \end{cases}$$

To cover its losses, the manufacturer purchases an insurance policy with an annual deductible of 2.

What is the mean of the manufacturer's annual losses not paid by the insurance policy?

Answer Let Y be the insurance payment amount. Then

$$Y = \begin{cases} 0, & \text{for } 0.6 < x \le 2 \\ X - 2, & x > 2 \end{cases}$$

Hence, the annual losses not paid is

$$X - Y = \begin{cases} X, & \text{for } 0.6 < x \le 2 \\ 2, & x > 2 \end{cases}$$

The expected amount of losses not paid by the policy is

$$\int_{0.6}^{2} xf(x)dx + \int_{2}^{\infty} 2f(x)dx = \int_{0.6}^{2} x \frac{2.5(0.6)^{2.5}}{x^{3.5}} dx + 2\int_{2}^{\infty} \frac{2.5(0.6)^{2.5}}{x^{3.5}} dx$$

$$= 2.5(0.6)^{2.5} \left[\frac{-x^{-1.5}}{1.5} \bigg|_{0.6}^{2} + 2 * \frac{-x^{-2.5}}{2.5} \bigg|_{2}^{\infty} \right]$$

$$= 2.5(0.6)^{2.5} \left(\frac{-2^{-1.5}}{1.5} + \frac{0.6^{-1.5}}{1.5} + 2\frac{2^{-2.5}}{2.5} \right) = 0.9343.$$

Question 17 *A tour operator has a bus that can accommodate 20 tourists. The operator knows that tourists may not show up, so he sells 21 tickets. The probability that an individual tourist will not show up is 0.02, independent of all other tourists.*

Each ticket costs 50 and is nonrefundable if a tourist does not show up. If a tourist shows up and a seat is not available, the tour operator has to pay 100 (ticket cost + penalty) to the tourist.

What is the expected revenue of the tour operator?

Answer The tour operator collects $50 \times 21 = 1050$ for the 21 tickets sold. If all 21 tourists show up, the tour operator will lose 100. The probability of losing 100 is $1(1 - 0.02)^{21} = 0.65$. Therefore, the expected revenue is $1050 - 100(0.65) = 985$.

Question 18 *Let X be the number of policies sold by an agent in a day. The moment generating function of X is*

$$M(t) = 0.45e^t + 0.35e^{2t} + 0.15e^{3t} + 0.05e^{4t}, \ for - \infty < t < \infty.$$

Calculate the standard deviation of X.

Answer It can be recognized that this MGF corresponds to a discrete random variable that takes values of 1, 2, 3, and 4 with probabilities 0.45, 0.35, 0.15, and 0.05, respectively.

This yields a mean of $1(0.45) + 2(0.35) + 3(0.15) + 4(0.05) = 1.8$, and second moment of $1^2(0.45) + 2^2(0.35) + 3^2(0.15) + 4^2(0.05) = 4$. The variance is $4 - 1.8^2 = 0.76$ and the standard deviation is $(0.76)^{0.5} = 0.87$.

Alternatively, using the relationship between MGF and moments of the random variable, we can take the first and second derivatives of the MGF and let $t = 0$. That gives us the same values of the first two moments as the previous approach. The same answer results.

Question 19 *The proportion X of yearly claims that exceed 200 is a random variable with density function*

$$f(x) = \begin{cases} 60x^3(1-x)^2, & 0 < x < 1 \\ 0, & \text{otherwise} \end{cases}$$

Calculate $\mathrm{Var}\left[\dfrac{X}{1-X}\right]$

Answer

$$E\left[\frac{X}{1-X}\right] = \int_0^1 \frac{x}{1-x} f(x) dx = \int_0^1 \frac{x}{1-x} 60x^3(1-x)^2 dx$$

$$= 60\int_0^1 x^4(1-x)^1 dx = 60\left(\frac{x^5}{5} - \frac{x^6}{6}\right)\Big|_0^1 = 60\left(\frac{1}{5} - \frac{1}{6}\right) = 2.$$

$$E\left[\left(\frac{X}{1-X}\right)^2\right] = \int_0^1 \left(\frac{x}{1-x}\right)^2 f(x) dx = \int_0^1 \left(\frac{x}{1-x}\right)^2 60x^3(1-x)^2 dx$$

$$= 60\int_0^1 x^5 dx = 60\left(\frac{x^6}{6}\right)\Big|_0^1 = 60\left(\frac{1}{6}\right) = 10.$$

Therefore $\mathrm{Var}\left[\dfrac{X}{1-X}\right] = 10 - 2^2 = 6.$

Question 20 *Automobile losses reported to an insurance company are independent and uniformly distributed between 0 and 20,000. The company covers each such loss subject to a deductible of 5000.*

Calculate the probability that the total payout on 200 reported losses is between 1,000,000 and 1,200,000.

Answer Because the number of payouts (including payouts of zero when the loss is below the deductible) is large, we apply the Central Limit Theorem and assume the total payout is normal distribution.

We need first to find the mean, second moment, and variance of the payout from a single loss.

$$\text{Mean} = \int_0^{5000} 0 * f(x)dx + \int_{5000}^{20000} (x - 5000)f(x)dx = 0$$

$$+ \frac{1}{20000} \int_{5000}^{20000} (x - 5000)dx$$

$$= \frac{1}{20000} \left. \frac{(x - 5000)^2}{2} \right|_{5000}^{20000} = 5,625.$$

$$\text{Second moment} = \int_0^{5000} 0^2 * f(x)dx + \int_{5000}^{20000} (x - 5000)^2 f(x)dx = 0$$

$$+ \frac{1}{20000} \int_{5000}^{20000} (x - 5000)^2 dx$$

$$= \frac{1}{20000} \left. \frac{(x - 5000)^3}{3} \right|_{5000}^{20000}$$

$$= 56,250,000.$$

Variance $= 56,250,000 - 5,625^2 = 24,609,375.$

The total payout from 200 losses has a mean of $5,625*200 = 1,125,000$ and a variance of $24,609,375*200 = 4,921,875,000$, thus standard deviation of 70,156.

Assuming normal distribution for total payout, denoted by S, and the probability sought after is

$$P(1,000,000 < S < 1,200,000)$$

$$= P\left(\frac{1,000,000 - 1,125,000}{70156} < Z < \frac{1,200,000 - 1,125,000}{70156} \right)$$

$$= P(-1.7817 < Z < 1.0690) = 0.8575 - (1 - 0.9626) = 0.8201.$$

In the last two lines, Z represents standard normal distribution and the probabilities are looked up from the table of standard normal distribution (available for examinees of exam P).

Question 21 *Two insurers provide bids on an insurance policy to a large company. The bids must be between 2000 and 2200. The company decides to accept the lower bid if the two bids differ by 20 or more. Otherwise, the company will consider the two bids further. Assume that the two bids are independent and are both uniformly distributed on the interval from 2000 to 2200.*
Determine the probability that the company considers the two bids further.

Answer

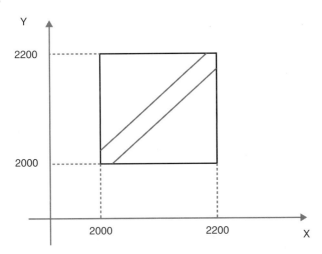

Because both variables, denoted X and Y, are uniformly distributed on the interval 2000 to 2200, the values of the bivariate (X,Y) are confined and uniformly distributed in the square in the figure. The probability that the company considers the two bids further is the ratio of the area confined by the red segments in the square to the area of the square.

The two red segments are defined by the equations $Y = X - 20$ and $Y = X + 20$. It is easier to compute the area of two congruent triangles, one above and one below the red segments.

The area of one such triangle is $\frac{1}{2} * 180 * 180$.

Therefore, the probability that the company considers the two bids further is equal to

$$1 - \frac{2*\frac{1}{2}*180*180}{200*200} = 1 - 0.9*0.9 = 0.19.$$

Question 22 *The waiting time for the first claim from a good driver and the waiting time for the first claim from a bad driver are independent and follow exponential distribution with means 6 years and 3 years, respectively.*

What is the probability that the first claim from a good driver will be filed within 3 years and the first claim from a bad driver will be filed within 2 years?

Answer For a good driver, the probability is $1 - e^{-\frac{3}{6}}$ and for a bad driver, the probability is $1 - e^{-\frac{2}{3}}$.

The probability in question is the product of both,

$$\left(1 - e^{-\frac{3}{6}}\right)\left(1 - e^{-\frac{2}{3}}\right) = 1 - e^{-\frac{1}{2}} - e^{-\frac{2}{3}} + e^{-\frac{7}{6}}.$$

Question 23 *Let X denote the proportion of employees at a large firm who will choose to be covered under the firm's medical plan, and let Y denote the proportion who will choose to be covered under both the firm's medical plan and dental plan.*

Suppose that for $0 \le y \le x \le 1$, X and Y have the joint cumulative distribution function:

$$F(x, y) = y\left(x^2 + xy - y^2\right).$$

Calculate the expected proportion of employees who will choose to be covered under both plans.

Answer The marginal cumulative distribution of Y is

$F_Y(y) = F(1, y) = y\left(1 + y - y^2\right) = y + y^2 - y^3$, thus its density function is $f_Y(y) = 1 + 2y - 3y^2$.

Therefore, the mean of Y is

$$E(Y) = \int_0^1 y\left(1 + 2y - 3y^2\right)dy = \frac{1}{2} + 2*\frac{1}{3} - 3*\frac{1}{4} = \frac{5}{12} = 0.417.$$

Question 24 *A device contains two circuits. The second circuit is a backup for the first, so the second is used only when the first has failed. The device fails when and only when the second circuit fails.*

Let X and Y be the times at which the first and second circuits fail, respectively. X and Y have joint probability density functions:

$$f(x,y) = \begin{cases} 6e^{-x}e^{-2y}, & 0 < x < y < \infty \\ 0, & \text{otherwise} \end{cases}.$$

What is the expected time at which the device fails?

Answer The marginal distribution of Y is

$$f_Y(y) = \int_0^y 6e^{-x}e^{-2y}dx = -6e^{-2y} * e^{-x}\big|_0^y = 6e^{-2y} - 6e^{-3y}.$$

The expected time at which the device fails is the expected value of Y (not $X + Y$), and its value is

$$E(Y) = \int_0^\infty y\left(6e^{-2y} - 6e^{-3y}\right)dy = 3\int_0^\infty y * 2e^{-2y}dy - 2\int_0^\infty y * 3e^{-3y}dy$$

$$= 3\left(\frac{1}{2}\right) - 2\left(\frac{1}{3}\right) = \frac{5}{6} = 0.83.$$

Question 25 *Random variables X and Y are uniformly distributed on the region bounded by the x and y axes, and the curve $y = 1 - x^2$.*
    Calculate E (XY).

Answer First we need to determine the density, a constant c, by solving the equation:

$$1 = \int_0^1 \int_0^{1-x^2} c \, dydx = \int_0^1 c(1 - x^2)dx = c\left(x - \frac{x^3}{3}\right)\Big|_0^1 = c\left(\frac{2}{3}\right).$$

Thus $c = \frac{3}{2} = 1.5$. Plugging in the following expression,

$$E(XY) = \int_0^1 \int_0^{1-x^2} cxy \, dydx = \int_0^1 \frac{1.5}{2} x y^2\big|_0^{1-x^2} dx$$

$$= \int_0^1 0.75x\left(1 - x^2\right)^2 dx = \int_0^1 0.75\left(x - 2x^3 + x^5\right)dx$$

$$= 0.75\left(\frac{x^2}{2} - \frac{2x^4}{4} + \frac{x^6}{6}\right)\Big|_0^1 = 0.125.$$

Question 26 *An insurance company sells automobile liability and collision insurance. Let X denote the percentage of liability policies that will be renewed at the end of their terms and Y the percentage of collision policies that will be renewed at the end of their terms. X and Y have the joint cumulative distribution function:*

$$F(x, y) = \frac{xy(x + y)}{2{,}000{,}000}, \text{ when } 0 \le x \le 100, 0 \le y \le 100.$$

Calculate Var (X).

Answer We first need to find the density function of X:

$$F_X(x) = F(x, 100) = \frac{100x(x + 100)}{2{,}000{,}000} = \frac{100x^2 + 10{,}000x}{2{,}000{,}000}.$$

Thus the density is $f_X(x) = \dfrac{x}{10{,}000} + \dfrac{1}{200}$.

$$E(X) = \int_0^{100} \left(\frac{x^2}{10{,}000} + \frac{x}{200} \right) dx = \left(\frac{x^3}{3 * 10{,}000} + \frac{x^2}{2 * 200} \right) \Bigg|_0^{100} = 58.33$$

$$E(X^2) = \int_0^{100} \left(\frac{x^3}{10{,}000} + \frac{x^2}{200} \right) dx = \left(\frac{x^4}{4 * 10{,}000} + \frac{x^3}{3 * 200} \right) \Bigg|_0^{100} = 4166.67$$

Therefore Var $(X) = 4166.67 - 58.33^2 = 764$.

Question 27 *An actuary determines that the annual number of tornadoes in counties P and Q are jointly distributed as in the table.*
  Calculate the conditional variance of the annual number of tornadoes in county Q, given that there are no tornadoes in county P.

		Annual number of tornadoes in county Q			
		0	1	2	3
Annual number of tornadoes in county P	0	0.12	0.06	0.05	0.02
	1	0.13	0.15	0.12	0.03
	2	0.05	0.15	0.10	0.02

Answer

$P(\text{no tornadoes in county P}, 0 \text{ tornadoes in county Q}) = 0.12$

$P(\text{no tornadoes in county P}, 1 \text{ tornado in county Q}) = 0.06$

$P(\text{no tornadoes in county P, 2 tornadoes in county Q}) = 0.05$

$P(\text{no tornadoes in county P, 3 tornadoes in county Q}) = 0.02$

Applying total probability formula, we have

$P(\text{no tornadoes in county P}) = 0.12 + 0.06 + 0.05 + 0.02 = 0.25.$

Conditional probabilities:

$P(0 \text{ tornadoes in county Q} \mid \text{no tornadoes in county P}) = \dfrac{0.12}{0.25} = \dfrac{12}{25}$

$P(1 \text{ tornadoes in county Q} \mid \text{no tornadoes in county P}) = \dfrac{0.06}{0.25} = \dfrac{6}{25}$

$P(2 \text{ tornadoes in county Q} \mid \text{no tornadoes in county P}) = \dfrac{0.05}{0.25} = \dfrac{5}{25}$

$P(3 \text{ tornadoes in county Q} \mid \text{no tornadoes in county P}) = \dfrac{0.02}{0.25} = \dfrac{2}{25}$

Hence

The conditional mean $= 0*\dfrac{12}{25} + 1*\dfrac{6}{25} + 2*\dfrac{5}{25} + 3*\dfrac{2}{25} = \dfrac{22}{25}$

The conditional 2nd moment $= 0^2*\dfrac{12}{25} + 1^2*\dfrac{6}{25} + 2^2*\dfrac{5}{25} + 3^2*\dfrac{2}{25} = \dfrac{44}{25}$

The conditional variance $= \dfrac{44}{25} - \left(\dfrac{22}{25}\right)^2 = 0.9856.$

Question 28 *New dental and medical plan options will be offered to state employees next year. An actuary uses the following density function to model the joint distribution of the proportion X of state employees who will choose Dental Option 1 and the proportion Y who will choose Medical Option 1 under the new plan options:*

$$f(x,y) = \begin{cases} 0.50, & \text{for } 0 < x < 0.5 \text{ and } 0 < y < 0.5 \\ 1.25, & \text{for } 0 < x < 0.5 \text{ and } 0.5 < y < 1 \\ 1.50, & \text{for } 0.5 < x < 1 \text{ and } 0 < y < 0.5 \\ 0.75, & \text{for } 0.5 < x < 1 \text{ and } 0.5 < y < 1 \end{cases}$$

Calculate Var $(Y|X = 0.75)$.

Answer First we need to find the marginal density at $X = 0.75$. It is equal to

$$f_X(0.75) = \int_0^1 f(0.75, y)\,dy = \int_0^{0.5} 1.5\,dy + \int_{0.5}^1 0.75\,dy = 1.125.$$

Now we can determine the conditional density of $Y|X = 0.75$:

$$f(y|X = 0.75) = \frac{f(0.75, y)}{f_X(0.75)} = \begin{cases} \dfrac{1.50}{1.125} = \dfrac{4}{3} & for\ 0 < y < 0.5 \\[3mm] \dfrac{0.75}{1.125} = \dfrac{2}{3} & for\ 0.5 < y < 1 \end{cases}.$$

Hence

$$E(Y|X = 0.75) = \int_0^1 y\,f(y\,|\,X = 0.75)\,dy$$

$$= \int_0^{0.5} y\left(\frac{4}{3}\right)dy + \int_{0.5}^1 y\left(\frac{2}{3}\right)dy$$

$$= \left(\frac{1}{8}\right)\left(\frac{4}{3}\right) + \left(\frac{3}{8}\right)\left(\frac{2}{3}\right) = \frac{5}{12}.$$

$$E(Y^2|X = 0.75) = \int_0^1 y^2 f(y|X = 0.75)\,dy$$

$$= \int_0^{0.5} y^2\left(\frac{4}{3}\right)dy + \int_{0.5}^1 y^2\left(\frac{2}{3}\right)dy$$

$$= \left(\frac{1}{24}\right)\left(\frac{4}{3}\right) + \left(\frac{7}{24}\right)\left(\frac{2}{3}\right) = \frac{18}{72}.$$

Therefore

$$\mathrm{Var}(Y|X = 0.75) = \left(\frac{18}{72}\right) - \left(\frac{5}{12}\right)^2 = \frac{11}{144} = 0.076.$$

Question 29 *An actuary models the lifetime of a device using the random variable $Y = 10X^{0.8}$, where X is an exponential random variable with mean 1 year.*

Determine the probability density function $f(y)$, for $y > 0$, of the random variable Y.

Answer For this type of questions, we always begin with linking the unknown cumulative distribution function to the other one we do know.

$$F(y) = P(Y \le y) = P\left(10X^{0.8} \le y\right) = P\left(X \le (0.1y)^{1.25}\right) = 1 - e^{-(0.1y)^{1.25}}.$$

Therefore, the density function is found by applying chain rule of differentiation,

$$f(y) = F'(y) = 1.25(0.1)(0.1y)^{0.25} e^{-(0.1y)^{1.25}}.$$

Question 30 *An insurance company issues* 1250 *vision care insurance policies. The number of claims filed by a policyholder under a vision care insurance policy during one year is a Poisson random variable with mean* 2. *Assume the number of claims filed by distinct policyholders are independent of one another.*

What is the approximate probability that there is a total of between 2450 *and* 2600 *claims during a one-year period?*

Answer A single policy has a mean and variance of two claims. For 1250 independent policies, the mean and variance of the total number of claims are both $2 \times 1250 = 2500$. The standard deviation is 50.

Using the Central Limit Theorem, the distribution of the total number of claims, denoted S, is approximately normal with mean 1250 and standard deviation 50. Thus

$$P(2450 < S < 2600) = P\left(\frac{2450 - 2500}{50} < \frac{S - 2500}{50} < \frac{2600 - 2500}{50}\right)$$
$$= P(-1 < Z < 2) = P(Z < 2) - P(Z < -1)$$
$$= 0.9772 - (1 - 0.8413) = 0.8185.$$

Where Z represents standard normal distribution.

2.12   Theory of Interest

Unlike most commercial transactions we encounter every day where we immediately receive the product that we pay for, an insurance transaction requires the buyer (policyholder) to make a payment (premium) first, and the seller (insurance company) makes a promise to pay only when specified events occur. That means a time lapse between the insurance company collecting premiums and paying the benefits, enabling the insurance company to invest the fund and earn a return. Financial transactions taking place at various spots of time require actuaries to understand and consider time value of money for the purpose of setting premiums and determining policy reserves.

Exam FM of SOA (called exam 2 of CAS) develops candidates' understanding of the fundamental concepts of financial mathematics, and trains candidates to learn how to apply these concepts in calculating present and accumulated values for various streams of cash flows as a basis for future use in reserving, valuation, pricing, asset/liability management, investment income, capital budgeting and valuing contingent cash flows.

Starting October 2022, Exam FM becomes a 2.5-hour exam with 30 multiple-choice questions.

Exam FM sample questions and answers

Here are some examples of how these ideas and techniques can be tested.

Q1 Present value of a sequence of payments, and of a single payment using nominal interest rate.

Q2 Use force of interest to find accumulated value.

Q3 Multiple nominal interest rates in determining accumulated value of one-time deposits.

Q4 Present values of multiple one-time deposits.

Q5 Exact match of liabilities and assets.

Q6 Accumulated values of multiple annuities immediately.

Q7 Present value of perpetuity with increasing payments for a fixed period of time.

Q8 Present value of annuity with payments forming geometric series.

Q9 Present values of annuities with different payment frequency.

Q10 Withdrawals of equal principals along with interests earned.

Q11 Present value of perpetuity equals accumulated value of annuity due.

Q12 Total interest on a loan with level payments.

Q13 Outstanding balance of a loan with payments forming a geometric series.

Q14 The decomposition of a loan payment into principal repaid and interest due.

Q15 Interest paid and principal repaid from different periods of a loan with level payments.

Q16 Equal total sums of all payments under different payment options.

Q17 Effective interest rate and force of interest over different periods of an agreement.

Q18 Price of a par value bond.

Q19 Price of a callable bond.

Q20 Price of a callable bond with unknown par value.

Q21 Present value of an annuity equals accumulated value of another annuity.

Q22 Spot interest rates used in pricing a bond.

Q23 Macaulay duration used in estimating percentage change of bond price.

Q24 Immunization, equal present values, and durations of assets and liabilities.

Q25 Compute spot rate from forward rates.

Q26 Macaulay durations immediately before and after a coupon payment.

Q27 Macaulay durations of annuities with different terms.

Q28 Concepts regarding immunization.

Q29 Zero-coupon bonds used in matching value and duration of assets and liabilities.

Q30 Zero-coupon bonds and coupon bonds are used in matching assets to liabilities.

Q31 Full immunization of a single liability using two assets of different terms.

Q32 Strategy with lowest cost to meet liabilities due at different times.

Question 1 *An investor's retirement account pays an annual nominal interest rate of 4.2%, convertible monthly. On January 1 of year y, the investor's account balance was X. The investor then deposited 100 at the end of every quarter. On May 1 of year (y + 10), the account balance was 1.9X.*
   Establish the equation that can be used to solve for X.

Answer The monthly interest rate is $\dfrac{0.042}{12} = 0.0035$. The quarterly rate is $1.0035^3 - 1 = 0.0105$. The investor makes 41 quarterly deposits, and the ending date is 124 months from the start. Using January 1 of year y as the comparison date produces the following equation:

$$X + \sum_{k=1}^{41} \frac{100}{1.0105^k} = \frac{1.9X}{1.0035^{124}}.$$

Question 2 *Ernie makes deposits of 100 at time 0, and X at time 3. The fund grows at a force of interest $\delta_t = \dfrac{t^2}{100}, t > 0$. The amount of interest earned from time 3 to time 6 is also X. Calculate X.*

Answer The accumulation function is $a(t) = \exp\left[\int_0^t (s^2/100)\,ds\right] = \exp\left(\dfrac{t^3}{300}\right)$

The accumulated value of 100 at time 3 is $100\exp\left(\dfrac{3^3}{300}\right) = 109.41743$.

The amount of interest earned from time 3 to time 6 equals the accumulated value of at time 6 minus the accumulated value at time 3. Thus

$$(109.41743 + X)\left[\frac{a(6)}{a(3)} - 1\right] = X$$

$$(109.41743 + X)\left[\frac{2.0544332}{1.0941743} - 1\right] = X.$$

Therefore $X = 784.61$.

Question 3 *Lucas opens a bank account with 1000 and lets it accumulate at an annual nominal interest rate of 6% convertible semiannually. Danielle also opens a bank account with 1000 at the same time as Lucas, but it grows at an annual nominal interest rate of 3% convertible monthly.*

For each account, interest is credited only at the end of each interest conversion period.

Calculate the number of months required for the amount in Lucas's account to be at least double the amount in Danielle's account.

Answer Let $n =$ years needed. The equation to solve is

$$1000\left(1 + \frac{0.06}{2}\right)^{2n} = 2(1000)\left(1 + \frac{0.03}{12}\right)^{12n}.$$

Take natural logarithm,

$$\ln 1000 + 2n\ln 1.03 = \ln 2 + \ln 1000 + 12n\ln 1.0025$$

$$n = 23.775.$$

This is 285.3 months. The next interest payment to Lucas is at a multiple of 6, which is 288 months.

Question 4 *A couple decides to save money for their child's first-year college tuition. The parents will deposit* 1700 *n months from today and another* 3400 2n *months from today. All deposits earn interest at a nominal annual rate of* 7.2%, *compounded monthly.*

Calculate the maximum integral value of n such that the parents will have accumulated at least 6500 *five years from today.*

Answer The monthly interest rate is $\dfrac{0.072}{12} = 0.006$. The amount of 6500 five years from today has present value of $6500(1.006)^{-60} = 4539.77$. The equation of value is $4539.77 = 1700(1.006)^{-n} + 3400(1.006)^{-2n}$.

Let $x = 1.006^{-n}$. Then solve the quadratic equation:

$$3400x^2 + 1700x - 4539.77 = 0.$$

We have $x = 0.93225$ (omit the negative root).
Then $1.006-^n = 0.9325$. Solve it and we get $n = 11.73$.
To ensure there is 6500 in five years, the deposits must be made earlier and thus the maximum integral value of n is 11.

Question 5 *Joe must pay liability of* 1,000 *due* 6 *months from now and another* 1,000 *due one year from now. There are two available investments:*

Bond I: A 6-month bond with face amount of 1,000, *an* 8% *nominal annual coupon rate convertible semiannually, and a* 6% *nominal annual yield rate convertible semiannually.*

Bond II: A 1-year bond with face amount of 1,000, *a* 5% *nominal annual coupon rate convertible semiannually, and a* 7% *nominal annual yield rate convertible semiannually.*

Calculate the amount of each bond that Joe should purchase to exactly match the liabilities.

Answer Because only Bond II provides a cash flow one year from now, it must be considered first. The bond provides 1025 one year from now and the number of units needed for this bond is $\dfrac{1000}{1025} = 0.97561$ to provide the required cash. This bond also provides $0.97561(25) = 24.39025$ six months from now. Thus Bond I must provide $1000 - 24.39025 = 975.60975$ six months from now. The bond provides 1040 per unit, hence the number of units needed is $\dfrac{975.60975}{1040} = 0.93809$.

Question 6 *To accumulate 8000 at the end of 3n years, deposits of 98 are made at the end of each of the first n years and 196 at the end of each of the next 2n years.*

The annual effective rate of interest is i. You are given $(1+i)^n = 2.0$. *Calculate i.*

Answer The equation of value is $98\, s_{\overline{3n}|} + 98\, s_{\overline{2n}|} = 8000$

That gives us

$$\frac{(1+i)^{3n} - 1}{i} + \frac{(1+i)^{2n} - 1}{i} = 81.63.$$

Since $(1+i)^n = 2.0$,

$$\frac{2^3 - 1}{i} + \frac{2^2 - 1}{i} = 81.63.$$

Finally $i = 0.1225 = 12.25\%$.

Question 7 *A perpetuity costs 77.1 and makes end-of-year payments. The perpetuity pays 1 at the end of year 2, 2 at the end of year 3, ..., n at the end of year n+1. After year n+1, the payments remain constant at n. The annual effective interest rate is 10.5%. Calculate n.*

Answer

$$77.1 = v(Ia)_{\overline{n}|} + \frac{nv^{n+1}}{i} = v\left[\frac{\ddot{a}_{\overline{n}|} - nv^n}{i}\right] + \frac{nv^{n+1}}{i}$$

$$= \frac{a_{\overline{n}|}}{i} - \frac{nv^{n+1}}{i} + \frac{nv^{n+1}}{i} = \frac{a_{\overline{n}|}}{i}$$

$$= \frac{1 - v^n}{i^2} = \frac{1 - v^n}{0.105^2}$$

Thus $1 - v^n = 0.85003$ and $1.105^{-n} = 0.14997$.

Therefore $n = -\dfrac{\ln(0.14997)}{\ln(1.105)} = 19.$

Note: To obtain the present value without remembering the formula for an increasing annuity, consider the payments as a perpetuity of 1 starting at time 2, a perpetuity of 1 starting at time 3, up to a perpetuity of 1 starting at time $n + 1$. The present value one year before the start of each perpetuity is $\dfrac{1}{i}$. The total present value is $\dfrac{1}{i}(v + v^2 + \ldots + v^n) = \dfrac{1}{i}a_{\overline{n}|}.$

Question 8 *Seth has two retirement benefit options.*
His first option is to receive a lump sum of 374,5000 at retirement.
His second option is to receive monthly payments for 25 years starting one month after retirement. For the first year, the amount of each monthly payment is 2000. For each subsequent year, the monthly payments are 2% more than the monthly payments from the previous year.
Using an annual nominal interest rate of 6%, compounded monthly, the present value of the second option is P. Determine the difference between P and the lump sum option amount.

Answer The accumulated value of the first year of payments is $2000s_{\overline{n}|} = 24{,}671.12$. This amount increases at 2% per year. The effective annual interest rate is $\left(1 + \dfrac{0.06}{12}\right)^{12} - 1 = 0.061678$. The present value is then

$$P = 24{,}671.12 \sum_{k=1}^{25} 1.02^{k-1} \left(1.061678\right)^{-k} = \frac{24{,}671.12}{1.02} \sum_{k=1}^{25} \left(\frac{1.02}{1.061678}\right)^{k}$$

$$= 24{,}187.37 \frac{0.960743 - 0.960743^{26}}{1 - 0.960743} = 374{,}444.$$

This is 56 less than the lump sum amount.

Question 9 *John finances his daughter's college education by making deposits into a fund earning interest at an annual effective rate of 8%. For 18 years he deposits X at the beginning of each month.*
In the 16th through the 19th years, he makes a withdrawal of 25,000 at the beginning of each year. The final withdrawal reduces the fund balance to zero. Calculate X.

Answer The effective monthly interest rate is $\left(1.08\right)^{-12} - 1 = 0.006434$.

The equation of value is $\dfrac{1}{1.08^{15}} 25{,}000\, \ddot{a}_{\overline{4}|.08} = X\ddot{a}_{\overline{216}|.006434}$.

Therefore $X = \dfrac{25{,}000\left(3.57710\right)}{3.17217\left(117.2790\right)} = 240.38.$

Question 10 *1000 is deposited into Fund X, which earns an annual effective rate of 6%. At the end of each year, the interest earned plus an additional 100 is withdrawn from the fund. At the end of the 10th year, the fund is depleted.*
The annual withdrawals of interest and principal are deposited into Fund Y, which earns an annual effective rate of 9%.
Calculate the accumulated value of Fund Y at the end of year 10.

Answer The interest earned and withdrawn is a decreasing annuity of 60, 54, 45, ..., 6. The series is deposited into Fund Y, along with the annual withdrawn principal of 100.

The accumulated value of Fund Y is:

$$6(Ds)_{\overline{10}|0.09} + 100\,s_{\overline{10}|0.09} = 6\left(\frac{10(1.09)^{10} - s_{\overline{10}|0.09}}{0.09}\right) + 100\,(15.19293)$$

$$= 565.38 + 1519.29 = 2084.67.$$

Question 11 *A man turns 40 today and wishes to provide supplement income of 3000 at the beginning of each month starting on his 65th birthday. Starting today, he makes monthly contributions of X to a fund for 25 years. The fund earns an annual nominal interest rate of 8% compounded monthly.*

On his 65th birthday, each 1000 of the fund will provide 9.65 of income at the beginning of each month starting immediately and continuing as long as he survives.

Calculate X.

Answer To receive 3000 per month at age 65, the fund must accumulate to

$$1000\left(\frac{3000}{9.65}\right) = 310,880.83.$$

The equation of value is $310,880.83 = X\,\ddot{s}_{\overline{300}|0.08/12} = 957.36657X$. Therefore $X = 324.72$.

Question 12 *Tanner takes out a loan today and repays the loan with 8-level annual payments, with the first payment due one year from today. The payments are calculated based on an annual effective interest rate of 4.75%. The principal portion of the 5th payment is 699.68. Calculate the total amount of interest paid on this loan.*

Answer Using the formula for a loan with level payments, the equation for annual payment P is

$$Pv^{8-5+1} = 699.68.\ So\ P = 842.39\ \text{and the total payments is}\ 8(842.39) = 6739.12$$

The principal portion of the first payment is $\dfrac{699.68}{1.0475^4} = 581.14$, and the interest portion of the first payment is $842.39 - 581.14 = 261.25$. Therefore, the loan amount is $\dfrac{261.25}{0.0475} = 5500$.

Finally, the total interest paid is $6739.12 - 5500 = 1239.12$.

Question 13 *A loan is amortized over five years with monthly payments at an annual nominal interest rate of 9% compounded monthly. The first payment is 1000 and is to be paid one month from the date of the loan. Each successive monthly payment will be 2% lower than the prior payment.*

Calculate the outstanding loan balance immediately after the 40th payment is made.

Answer Monthly payment at time t is $1000(0.98)^{t-1}$.

Because the loan amount is unknown, the outstanding balance must be calculated prospectively. The value at time 40 months is the present value of payments from time 41 to time 60.

Using $v = \dfrac{1}{1.0075}$, $OB_{40} = 1000\left[0.98^{40}v^1 + 0.98^{41}v^2 + \ldots + 0.98^{59}v^{20}\right]$

$$= 1000\frac{0.98^{40}v - 0.98^{60}v^{21}}{1 - 0.98v} = 1000\frac{0.44238 - 0.25434}{1 - 0.97270} = 6888.$$

Question 14 *Ron is repaying a loan with payments of 1 at the end of each year for n years. The annual effective interest rate on the loan is i. The amount of interest paid in year t + 1 plus the amount of principal repaid in year t + 1 equals X. Formulate an equation in order to determine X.*

Answer Year t interest is $ia_{\overline{n-t+1}|i} = 1 - v^{n-t+1}$.

Year $t+1$ principal repaid is $1 - \left(1 - v^{n-t}\right) = v^{n-t}$.

Therefore $X = 1 - v^{n-t+1} + v^{n-t} = 1 + v^{n-t}\left(1 - v\right) = 1 + v^{n-t}d$.

Question 15 *A loan of X is repaid with level annual payments at the end of each year for 10 years. You are given:*

(i) The interest paid in the first year is 3600; and
(ii) The principal repaid in the 6th year is 4871.

Calculate X.

Answer Let P be the annual payment. We have $P\left(1 - v^{10}\right) = 3600$, and $P v^{10-6+1} = 4871$.

Thus $\dfrac{1 - v^{10}}{v^5} = \dfrac{3600}{4871}$, or $1 - v^{10} = 0.739068v^5$.

Solving a quadratic equation yields $v^5 = 0.69656$.

Hence $i = 0.69596^{-1/5} - 1 = 0.075$.

Therefore $X = \dfrac{P(1 - v^{10})}{i} = \dfrac{3600}{0.075} = 48,000$.

Question 16 *A 10-year loan of 2000 is to be repaid with payments at the end of each year. It can be repaid under the following two options:*

(1) Equal annual payments at an annual effective interest rate of 8.07%.
(2) Installments of 200 each year plus interest on the unpaid balance at an annual effective interest rate of i.

The sum of the payments under option (i) equals the sum of the payments under option (ii).
Calculate i.

Answer

Option 1: $2000 = Pa_{\overline{10}|0.0807}$, so that $P = 299$, and the total payment is 2990.

Option 2: Interest paid must be $2990 - 2000 = 990$.

$990 = i[2000 + 1800 + 1600 + \ldots + 200] = 11000i$.

Hence $i = 0.09 = 9\%$.

Question 17 *A bank agrees to lend 10,000 now and X three years from now in exchange for a single repayment of 75,000 at the end of 10 years. The bank charges interest at an annual effective rate of 6% for the first 5 years and at a force of interest $\delta_t = \dfrac{1}{t+1}$ for $t \geq 5$. Calculate X.*

Answer

$$\left[10,000(1.06)^5 + X(1.06)^2\right] e^{\int_5^{10} \frac{1}{t+1}\,dt} = 75,000.$$

$$(13,382.26 + 1.1236X) * \frac{11}{6} = 75,000.$$

$$X = 24,498.78.$$

Question 18 *A 1000 par value 20-year bond sells for P and yields a nominal interest rate of 10% convertible semiannually. The bond has 9% coupons payable semiannually and a redemption value of 1200. Calculate P.*

Answer The coupons are $\dfrac{1000(0.09)}{2} = 45$ each.

The present value of the coupons and redemption value at 5% per half year is

$$P = 45a_{\overline{40}|0.05} + 1200(1.05)^{-40} = 942.61.$$

Question 19 *An investor purchases a 10-year callable bond with face amount of 1000 for price P. The bond has an annual nominal coupon rate of 10% paid semiannually.*

The bond may be called at par by the issuer on every other coupon payment date, beginning with the second coupon payment date.

The investor earns at least an annual nominal yield of 12% compounded semi-annually regardless of when the bond is redeemed.

Calculate the largest possible value of P.

Answer For a bond bought at discount (coupon rate is less than yield), the maximum price the investor is willing to pay is the smallest price possible, which occurs at the latest possible redemption date.

$$P = 50a_{\overline{20}|0.06} + 1000(1.06)^{-20} = 885.30.$$

Question 20 *Sue purchased a 10-year par value bond with an annual nominal coupon rate of 4% payable semi-annually at a price of 1021.50. The bond can be called at par value X on any coupon date starting at the end of year 5. The lowest yield rate that Sue can possibly receive is an annual nominal rate of 6% convertible semi-annually. Calculate X.*

Answer The bond is sold at a discount (coupon rate is less than yield rate), so the lowest yield rate is calculated based on a call at the latest possible redemption date. The par value X is also the redemption value because the bond is a par value bond. Therefore

$$1021.50 = 0.02Xa_{\overline{20}|0.03} + Xv^{20} = 0.851225X$$

$$X = 1200.$$

Question 21 *Happy and financially astute parents decide at the birth of their daughter that they will need to provide 50,000 at each of their daughter's 18th, 19th, 20th, and 21st birthdays to fund her college education. They plan to contribute X at each of their daughter's 1st through 17th birthdays to fund the four 50,000 withdrawals. They anticipate earning a constant 5% annual effective interest rate on their contributions.*

Let $v = \dfrac{1}{1.05}$. Determine the equation of value used to calculate X.

Answer The present value of the four 50,000 withdrawals on the 18th birthday is $50,000(1 + v + v^2 + v^3)$.

The accumulated value of the 17 contributions on the 18th birthday is $X\left[1.05^{17} + 1.05^{16} + \ldots + 1.05\right]$.

Thus the equation is

$$X\left[1.05^{17} + 1.05^{16} + \ldots + 1.05\right] = 50,000(1 + v + v^2 + v^3).$$

Question 22 *You are given the following information with respect to a bond:*
Par value: 1000
Term to maturity: 3 years
Annual coupon rate: 6% payable annually
You are also given the at the 1, 2, and 3 year annual spot interest rates are 7%, 8%, and 9%, respectively.
The bond is sold at a price equal to its value.
Calculate the annual effective yield rate for the bond i .

Answer The coupon payment is 60, payable at the end of each year. Principal of 1000 is repaid at the end of the third year.

Using spot rates, the value of the bond is:

$$\frac{60}{1.07} + \frac{60}{1.08^2} + \frac{1060}{1.09^3} = 926.03.$$

The annual effective yield rate is the solution to

$$926.03 = 60a_{\overline{3}|i} + 1000(1 + i)^{-3}. \text{ The solution is 8.9\% using a calculator.}$$

Question 23 *A 20-year bond priced to have an annual effective yield of 10% has a Macaulay duration of 11. Immediately after the bond is priced, the market yield rate increases by 0.25%. The bond's approximate percentage price change, using a first-order modified approximation, is X. Calculate X.*

Answer The modified duration is $\dfrac{11}{1 + 0.1} = 10$. Let $P(i)$ be the bond price at yield rate i .

Then $P(0.1025) \approx P(.10)\left[1 - (0.1025 - 0.10)(10)\right] = 0.975\, P(10).$

Therefore, the approximate percentage price change is $100(0.975 - 1) = -2.5\%$.

Question 24 *A liability consists of a series of 15 annual payments of 35,000 with the first payment to be made one year from now.*

The assets available to immunize this liability are five-year and ten-year zero-coupon bonds.

The annual effective interest rate used to value the assets and the liability is 6.2%. The liability has the same present value and duration as the asset portfolio.

Calculate the amount invested in the five-year zero-coupon bonds.

Answer First, the present value of the liability is $PV = 35,000 a_{\overline{15}|0.062} = 335,530.30$.

The duration of the liability is

$$d = \frac{35,000v + 2(35,000)v^2 + \ldots + 15(35,000)v^{15}}{335,530.30} = \frac{2,312,521.95}{335,530.30} = 6.89214.$$

Let X denote the amount invested in the 5-year bond. Then

$$\frac{X}{335,530.30}(5) + \left(1 - \frac{X}{335,530.30}\right)(10) = 6.89214.$$

Therefore $X = 208,556$.

Question 25 *The one-year forward rates, deferred t years, are estimated to be:*

Year (t)	0	1	2	3	4
Forward rate	4%	6%	8%	10%	12%

Calculate the spot rate for a zero-coupon bond maturing three years from now.

Answer $s_1 = {}_1f_0 = 0.04$

$${}_1f_1 = 0.06 = \frac{(1 + s_2)^2}{1 + s_1} - 1, \text{ hence } s_2 = \left((1.06)(1.04)\right)^{0.5} - 1 = 0.04995.$$

$${}_1f_2 = 0.08 = \frac{(1 + s_3)^3}{(1 + s_2)^2} - 1, \text{ hence } s_3 = \left((1.08)(1.04995)^2\right)^{1/3} - 1 = 0.05987 = 6\%.$$

Question 26 *Sam buys an 8-year, 5000 par bond with an annual coupon rate of 5%, paid annually. The bond sells for 5000. Let d_1 be the Macaulay duration just before the first coupon is paid. Let d_2 be the Macaulay duration just after the first coupon is paid. Calculate $\dfrac{d_1}{d_2}$.*

Answer Let d_0 be the Macaulay duration at time zero.

$$d_0 = \ddot{a}_{\overline{8}|0.05} = 6.7864$$

$$d_1 = d_0 - 1 = 5.7864$$

$$d_2 = \ddot{a}_{\overline{7}|0.05} = 6.0757$$

Therefore $\dfrac{d_1}{d_2} = \dfrac{5.7864}{6.0757} = 0.9524.$

This solution employs the fact that when a coupon bond sells at par the duration equals the present value of an annuity-due. For the duration just before the first coupon, the cash follows are the same as for the original bond, but all occur one year sooner. Hence, the duration is one year less.

Question 27 *Annuity A pays 1 at the beginning of each year for three years. Annuity B pays 1 at the beginning of each year for four years. The Macaulay duration of Annuity A at the time of purchase is 0.93. Both annuities offer the same yield rate. Calculate the Macaulay duration of Annuity B at the time of purchase.*

Answer The Macaulay duration of Annuity A is $0.93 = \dfrac{0(1) + 1(v) + 2(v^2)}{1 + v + v^2} = \dfrac{v + 2v^2}{1 + v + v^2}$, solving the quadratic equation leads to $v = 0.9$.

The Macaulay duration of Annuity B is $\dfrac{0(1) + 1(v) + 2(v^2) + 3(v^3)}{1 + v + v^2 + v^3} = 1.369.$

Question 28 *Which of the following statements regarding immunization are true?*

I. *If long-term interest rates are lower than short-term rates, the need for immunization is reduced.*
II. *Either Macaulay or modified duration can be used to develop an immunization strategy.*
III. *Both processes of matching the present values of the flows or the flows themselves will produce exact matching.*

Answer

I.   False. The yield curve structure is not relevant.

II. True.

III. False. Matching the present values is not sufficient when interest rates change.

Question 29 *A company has liabilities of* 402.11 *due at the end of each of the next three years. The company will invest* 1000 *today to fund these payouts. The only investments available are one-year and three-year zero-coupon bonds, and the yield curve is flat at a* 10% *annual effective rate. The company wishes to match the duration of its assets to the duration of its liabilities.*

Determine how much the company should invest in each bond.

Answer The present value of the liabilities is 1000, so that requirement is met.

The duration of the liabilities is

$$\frac{402.11\left[1.1^{-1} + 2(1.1)^{-2} + 3(1.1)^{-3}\right]}{1000} = 1.9365.$$

Let X be the investment in the one-year bond. The duration of a zero-coupon bond is its term. The duration of the two bonds is then

$$\frac{\left[X(1) + (1000 - X)(3)\right]}{1000} = 3 - 0.002X.$$

Setting this equal to 1.9365 and solving the equation yields $X = 531.75$.

Question 30 *An insurance company must pay liabilities of* 99 *at the end of one year,* 102 *at the end of two years, and* 100 *at the end of three years. The only investments available are the following three bonds. Bond A and Bond C are annual coupon bonds. Bond B is a zero-coupon bond.*

Bond	Maturity (years)	Yield to maturity (annualized)	Coupon rate
A	1	6%	7%
B	2	7%	0
C	3	9%	5%

All three bonds have a par value of 100 *and will be redeemed at par.*

Calculate the number of units on Bond A that must be purchased to match the liabilities exactly.

Answer Let N be the number of units bought of the bond as indicated by the subscript.

$N_C * (105) = 100$, so that $N_C = 0.9524$.

$N_B{}^*(100) = 102 - (0.9524)(5)$, So that $N_B = 0.9724$.

$N_A{}^*(107) = 99 - (0.9524)(5)$, So that $N_A = 0.8807$.

Question 31 *Aakash has a liability of 6000 due in four years. This liability will be met with payments of A in two years and B in six years. Aakash is employing a full immunization strategy using an annual effective interest rate of 5%. Calculate $|A - B|$.*

Answer Set up the equations with the end of four years as the evaluation point:

$$A(1.05)^2 + B(1.05)^{-2} = 6000$$

$$2A(1.05)^2 - 2B(1.05)^{-2} = 0$$

Solving the equations yields:

$$A = 2721.09, B = 3307.50.$$

Therefore $|A - B| = 586.41$.

Question 32 *A bank accepts a 20,000 deposit from a customer on which it guarantees to pay an annual effective interest rate of 10% for 2 years. The customer needs to withdraw half of the accumulated value at the end of the first year. The customer will withdraw the remaining value at the end of the second year.*

The bank has the following investment options available, which may be purchased in any quantity.

Bond H: A one-year zero-coupon bond yielding 10% annually.

Bond I: A two-year zero-coupon bond yielding 11% annually.

Bond J: A two-year bond that sells at par with 12% annual coupons.

Any portion of the 20,000 deposit that is not needed to be invested in bonds is retained by the bank as profit.

Determine the investment strategy that produces the highest profit for the bank and is guaranteed to meet the customer's withdrawal needs.

Answer The correct answer is the lowest cost portfolio that provides for 11,000 at the end of year one and 12,100 at the end of year two.

Let H, I, and J be the face amount of each bond purchased.

The year one payment can be exactly matched with $H + 0.12 J = 11,000$.

The year two payment can be exactly matched with $I + 1.12 J = 12,00$.

The cost of the three bonds purchased is $\dfrac{H}{1.1} + \dfrac{I}{1.2321} + J$.

The cost is to be minimized under the two constraints. Substituting for H and I gives $\dfrac{11,000 - 0.12J}{1.1} + \dfrac{12,100 - 1.12J}{1.2321} + J = 19,820 - 0.0181J$.

This is minimized by purchasing the largest possible amount of J.

So we have $J = \dfrac{12,100}{1.12} = 10,803.57$. Then $H = 11,000 - 0.12\,(10,803.57) = 9703.57$, and the cost of Bond H is $\dfrac{9703.57}{1.1} = 8,821.43$.

2.13   Economics, Finance, and Accounting

Actuaries examine and try to understand economic phenomena, so knowledge of economics is helpful to the job. Actually, expected utility theory of economics is fundamental to understanding why insurance exists in the first place. One of the actuaries' key job functions is to make sure the insurance company has enough assets to cover its liabilities. Hence knowledge of corporation finance and accounting helps an actuary to appreciate the importance of her own job, and is perhaps instrumental to transition to management function.

Given the already heavy loads of actuarial examinations, economics, finance, and accounting are not subjects of any actuarial examination. Instead, students who intend to pursue actuarial careers are expected to take college courses in the subjects and earn a minimum grade in order to fulfill the requirements regarding these subjects. This "Validation by Educational Experience" is part of associateship requirements of both SOA and CAS.

Microeconomics

Microeconomics focuses on the role of individual firms and groups of firms with national and international economies. Key ideas of microeconomics are the demand and supply for individual goods and services, their trading and patterns of pricing, market equilibrium, and ideas such as the concepts of a monopoly, where one firm dominates the market, and an oligopoly, where a small number of firms dominate a national or global market.

In order to be recognized as VEE course, a college course on microeconomics is expected to address most of the topics:

- Explain the concept of utility and how rational utility maximizing agencies make consumption choices.
- Explain the elasticity of supply and demand and the effects on a market of the different levels of elasticity.

- Explain the interaction between supply and demand and the way in which equilibrium market prices are achieved.
- Explain various pricing strategies that can be used by firms.
- Explain the core economic concepts involved in choices made by businesses with respect to short-run and long-run investment and production choices.
- Explain competitive markets and how they operate.
- Explain profitability in markets with imperfect competition.

Macroeconomics

Macroeconomics deals with aggregate economic factors such as total national income and output, employment, balance of payments, rates of inflation, and the business cycle. One of the key ideas of macroeconomics is that of a gross national product: the total value of goods and services produced in an economy during a specified period of time.

A VEE course on this subject should cover most of the topics:

- Explain basic macroeconomic measures (e.g., GDP) used to compare the economies of countries.
- Describe the structure of public finances for an industrialized country.
- Explain the effect of fiscal and monetary policy on the economy, including the effect on financial markets.
- Explain the role of international trade, exchange rates, and the balance of payments in the economy.
- Explain the effect of savings and consumption rates on the economy.
- Explain the major factors affecting the level of interest rates, the rate of inflation, the exchange rate, the level of employment, and the rate of growth for an industrialized country.
- Describe the function of money in the economy.
- Explain the relationship between money and interest rates.
- Explain how macroeconomic policies affect businesses.

Accounting

Topics that should be covered in a VEE course on accounting include:

- Describe the basic principles of corporate taxation and the taxation of investments held by institutions.
- Explain why companies are required to produce annual reports.
- Explain fundamental accounting concepts and terms and describe the main sources of accounting regulation.
- Explain the structure and purpose of the income statement, balance sheet, and statement of cash flows and the interactions between them.
- Construct simple statements of financial position and profit or loss.
- Calculate and interpret financial ratios.

Corporate finance

Actuaries are expected to understand the following topics in finance:

- Explain the characteristics of various forms of equity capital from the point of view of the issuer and the investor.
- Explain the characteristics of various forms of long-term debt capital from the point of view of the issuer and the investor.
- Explain the characteristics of various forms of short- and medium-term financing from the point of view of the issuer and the investor.
- Calculate weighted-average cost of capital.
- Explain the main methods of capital budgeting.
- Calculate a project's investment return.

Exam P, Exam FM, VEE topics of economics, finance, and accounting are common requirements toward associateship of both SOA and CAS. Due to different risks in the core businesses of life insurance and property-casualty insurance, SOA and CAS differ on required skillsets of their members, although SOA has been trying to close the gap in recent years.

2.14 Actuarial Mathematics

Actuaries must master mathematical and statistical models that are employed to represent and explain events of interest, such as loss frequency, claim severity, and aggregate claims. Because losses and claims are not certain, models for contingent payments are also important.

SOA requires candidates to take and pass exam FAM (Fundamentals of Actuarial Mathematics) and any one of two exams: ASTAM (Advanced Short-Term Actuarial Mathematics) and ALTAM (Advanced Long-Term Actuarial Mathematics). VEE in Statistics provides necessary foundations in statistics for studying actuarial mathematics.

VEE mathematical statistics

Students usually earn the credits by taking university coursework or an online course.

Such a course should be calculus-based and cover all the topics that follow.

- Explain the concepts of random sampling, statistical inference and sampling distribution, and state and use basic sampling distributions.
- Describe the main methods of estimation and the main properties of estimators and apply them. Methods include matching moments, percentile matching, and maximum likelihood, and properties include bias, variance, mean squared error, consistency, efficiency, and UMVUE.
- Construct confidence intervals for unknown parameters, including the mean, differences of two means, variances, and proportions.

- Test hypotheses. Concepts to be covered include Neyman-Pearson lemma, significance and power, likelihood ratio test, and information criteria. Tests should include mean, variance, contingency tables, and goodness-of-fit.

Exam FAM topics

This exam trains candidates in acquiring fundamentals in models commonly used for short-term and long-term insurance coverage. It is a three-and-half-hour exam consisting of 40 multiple-choice questions.

The March 2023 syllabi of exam FAM describes learning outcomes and their weights in the exam.

1. Topic: Insurance and Reinsurance Coverages (7.5%–12.5%)
2. Topic: Severity, Frequency, and Aggregate Models (12.5%–15%)
3. Topic: Parametric and Nonparametric Estimation (5%–10%)
4. Topic: Introduction to Credibility (2.5%–5%)
5. Topic: Pricing and Reserving for Short-Term Insurance Coverages (7.5%–12.5%)
6. Topic: Option Pricing Fundamentals (2.5%–7.5%)
7. Topic: Insurance Coverages and Retirement Financial Security Programs (2.5%–7.5%)
8. Topic: Mortality Models (7.5%–12.5%)
9. Topic: Parametric and Nonparametric Estimation (5%–10%)
10. Topic: Present Value Random Variables for Long-Term Insurance Coverages (10%–15%)
11. Topic: Premium and Policy Value Calculation for Long-Term Insurance Coverages (15%–20%)

Exam FAM sample questions and solutions

The true exam questions are all multiple choice. Choices in the following sample questions have been omitted.

Question 1 *You are given:*

Number of claims	Probability	Claim size	Probability
0	1/5		
1	3/5	25	1/3
		150	2/3
2	1/5	50	2/3
		200	1/3

Claim sizes are independent.
Calculate the variance of the aggregate losses.

Answer First we obtain the distribution of aggregate losses:

Value	Probability
0	1/5
25	(3/5)(1/3) = 1/5
100	(1/5)(2/3)(2/3) = 4/45
150	(3/5)(2/3) = 2/5
250	(1/5)(2)(2/3)(1/3) = 4/45
400	(1/5)(1/3)(1/3) = 1/45

$$\text{Mean} = \left(\frac{1}{5}\right)(0) + \left(\frac{1}{5}\right)(25) + \left(\frac{4}{45}\right)(100) + \left(\frac{2}{5}\right)(150) + \left(\frac{4}{45}\right)(250)$$

$$+ \left(\frac{1}{45}\right)(400) = 105.$$

$$\text{Var} = \left(\frac{1}{5}\right)(0^2) + \left(\frac{1}{5}\right)(25^2) + \left(\frac{4}{45}\right)(100^2) + \left(\frac{2}{5}\right)(150^2) + \left(\frac{4}{45}\right)(250^2)$$

$$+ \left(\frac{1}{45}\right)(400^2) - 105^2 = 8100.$$

Question 2 *A towing company provides all towing services to members of the City Automobile Club. You are given:*

Towing distance	Towing cost	Frequency
0 – 9.99 miles	80	50%
10 – 29.99 miles	100	40%
30+ miles	160	10%

(i) *The automobile owner must pay 10% of the cost and the remainder is paid by the City Automobile Club.*

(ii) *The number of towing has a Poisson distribution with a mean of 1000 per year.*

(iii) *The number of towing and the cost of individual towing are all mutually independent.*

Calculate the probability that the club pays more than 90,000 in any year using the normal approximation for the distribution of aggregate towing costs.

Answer First we restate the table to be the club's cost, after the 10% payment by the auto owner:

Towing cost	Probability
72	50%
90	40%
144	10%

Then $E(X) = (0.5)(72) + (0.4)(90) + (0.1)(144) = 86.4$.

$$E(X^2) = (0.5)(72^2) + (0.4)(90^2) + (0.1)(144^2) = 7905.6.$$

$$\text{Var}(X) = 7905.6 - 86.4^2 = 440.64.$$

The frequency is Poisson, so $E(N) = \text{Var}(N) = 1000$, and

$$E(S) = E(N)E(X) = 86,400,$$

$$\text{Var}(S) = E(N)\text{Var}(X) + \text{Var}(N)\left[E(X)\right]^2$$
$$= 1000(440.64) + 1000(86.4^2) = 7,905,600.$$

$$\Pr(S > 90,000) = \Pr\left(\frac{S - E(S)}{\sqrt{\text{Var}(S)}} > \frac{90,000 - 86,400}{\sqrt{7,905,000}}\right)$$
$$= \Pr(Z > 1.28) = 1 - \Phi(1.28) = 0.10.$$

Question 3 *You are given:*

(i) *Losses follow an exponential distribution with the same mean in all years.*
(ii) *The loss elimination ratio this year is 70%.*
(iii) *The ordinary deductible for the coming year is 4/3 of the current deductible.*

Calculate the loss elimination ratio for the coming year.

Answer

$$\text{LER} = \frac{E(X \wedge d)}{E(X)} = \frac{\theta\left(1 - e^{-\frac{d}{\theta}}\right)}{\theta} = 1 - e^{-\frac{d}{\theta}}.$$

Last year $1 - e^{-\frac{d}{\theta}} = 0.70$, thus $d = -\theta \ln(0.30)$.

The coming year

$$d_{new} = \frac{4}{3}d = -\frac{4}{3}\theta \ln(0.30) \text{ thus} \frac{d_{new}}{\theta} = -\frac{4}{3}\ln(0.30) = 1.605297$$

$$\text{New LER} = 1 - e^{-\frac{d_{new}}{\theta}} = 1 - e^{-1.605297} = 0.80.$$

Question 4 *X is a discrete random variable with a probability function that is a member of the $(a,b,0)$ class of distributions.*
  You are given:

(i) $P(X = 0) = P(X = 1) = 0.25$

(ii) $P(X = 2) = 0.1875.$

  Calculate $P(X = 3)$.

Answer

$$p_k = \left(a + \frac{b}{k} \right) p_{k-1}$$

We have $0.25 = \left(a + \frac{b}{1} \right)(0.25)$, thus $a = 1 - b$.

And $0.1875 = \left(a + \frac{b}{2} \right)(0.25)$, thus $0.1875 = (1 - 0.5b)(0.25)$, hence

$b = 0.5$, $a = 0.5$.

Therefore $p_3 = \left(0.5 + \frac{0.5}{3} \right)(0.1875) = 0.125.$

Question 5 *The unlimited severity distribution for claim amounts under an auto liability insurance policy is given by the cumulative distribution:*

$$F(x) = 1 - 0.8e^{-0.02x} - 0.2e^{-0.001x}, x \geq 0.$$

The insurance policy pays amounts up to a limit of 1000 per claim.
Calculate the expected payment under this policy for one claim.

Answer

Limited expected value $= \int_0^{1000} \left[1 - F(x) \right] dx$

$$= \int_0^{1000} \left(0.8e^{-0.02x} + 0.2e^{-0.001x} \right) dx$$

$$= (-40e^{-0.02x} - 200e^{-0.001x}) \, |_0^{1000}$$

$$= 166.424.$$

Question 6 *You are given:*

(i) $S_0(t) = \left(1 - \dfrac{t}{\omega} \right)^{\frac{1}{4}}$, *for* $0 \le t \le \omega$.

(ii) $\mu_{65} = \dfrac{1}{180}$.

Calculate e_{106}, the curtate expectation of life at age 106.

Answer

Since $S_0(t) = \left(1 - \dfrac{t}{\omega} \right)^{\frac{1}{4}} = 1 - F_0(t)$, we have $\ln S_0(t) = \dfrac{1}{4} \ln \left[\dfrac{\omega - t}{\omega} \right]$.

Then $\mu_t = -\dfrac{d}{dt} \ln S_0(t) = \dfrac{1}{4} \dfrac{1}{\omega - t}$, and

$$\mu_{65} = \frac{1}{4} \frac{1}{\omega - 65} = \frac{1}{180}, \text{ thus } \omega = 110.$$

$$e_{106} = \sum_{t=1}^{3} {}_t p_{106} \text{ since } {}_4 p_{106} = 0.$$

$${}_t p_{106} = \frac{S_0(106 + t)}{S_0(106)} = \frac{\left(1 - \dfrac{106 + t}{110} \right)^{\frac{1}{4}}}{\left(1 - \dfrac{106}{110} \right)^{\frac{1}{4}}} = \left(\frac{4 - t}{4} \right)^{1/4}.$$

Finally

$$e_{106} = \sum_{t=1}^{3} {}_t p_{106} = \frac{1}{4^{0.25}} \left[3^{0.25} + 2^{0.25} + 1^{0.25} \right] = 2.4786.$$

Question 7 *You are given that mortality follows Makeham's Law with the following parameters:*

(i) $A = 0.004$

(ii) $B = 0.00003$

(iii) $c = 1.1$

Let L_{15} be the random variable representing the number of lives alive at the end of 15 years if there are 10,000 lives age 50 at time 0.

Calculate $Var[L_{15}]$.

Answer The distribution is binomial with 10,000 trials.

$$Var(L_{15}) = npq = 10,000 \left({}_{15}p_{50}\right)\left({}_{15}q_{50}\right)$$

where

$${}_{15}p_{50} = e^{\left[-A*15 - \frac{B}{lnc}*c^{50}\left(c^{15} - 1\right)\right]} = 0.837445$$

$${}_{15}q_{50} = 1 - {}_{15}p_{50} = 0.162555.$$

Therefore $Var(L_{15}) = 10,000 \left(0.837445\right)\left(0.162555\right) = 1361.3.$

Question 8 *You are given:*

(i) *An excerpt from a select and ultimate life table with a select period of 2 years:*

x	$l_{[x]}$	$l_{[x]+1}$	l_{x+2}	$x + 2$
50	99,000	96,000	93,000	52
51	97,000	93,000	89,000	53
52	93,000	88,000	83,000	54
53	90,000	84,000	78,000	55

(ii) *Deaths are uniformly distributed over each year of age.*

Calculate $10,000 \,{}_{2.2}q_{[51]+0.5}.$

Answer

$$10,000 \,{}_{2.2}q_{[51]+0.5} = 10,000 \frac{l_{[51]+0.5} - l_{53.7}}{l_{[51]+0.5}},$$

where

$$l_{[51]+0.5} = 0.5l_{[51]} + 0.5l_{[51]+1} = 0.5(97,000) + 0.5(93,000) = 95,000$$

$$l_{53.7} = 0.3l_{53} + 0.7l_{54} = 0.3(89,000) + 0.7(83,000) = 84,800$$

$$10,000 \, _{2.2}q_{[51]+0.5} = 10,000\frac{l_{[51]+0.5} - l_{53.7}}{l_{[51]+0.5}} = 10,000\frac{95,000 - 84,800}{95,000} = 1074.$$

Question 9 *For a special whole life insurance policy issued on (40), you are given:*

(i) *Death benefits are payable at the end of the year of death.*

(ii) *The amount of benefit is 2 if death occurs within the first 20 years and is 1 thereafter.*

(iii) *Z is the present value random variable for the payments under this insurance.*

(iv) *i = 0.03.*

x	A_x	$_{20}E_x$
40	0.36987	0.51276
60	0.62567	0.17878

(v) $E(Z^2) = 0.24954.$

Calculate the standard deviation of Z.

Answer

$$E(Z) = 2 * A_{40} - {}_{20}E_{40}A_{60} = 2(0.36987) - (0.51276)(0.62567) = 0.41892.$$

$E(Z^2) = 0.24954$ is given, so that

$$Var(Z) = 0.24954 - 0.41892^2 = 0.07405$$

$$SD(Z) = (0.07405)^{0.5} = 0.27212.$$

Question 10 *For a group of 100 lives age x with independent future lifetimes, you are given:*

(i) *Each life is to be paid 1 at the beginning of each year, if alive.*

(ii) *$A_x = 0.45$.*

(iii) $^2A_x = 0.22$.

(iv) $i = 0.05$.

(v) *Y is the present value random variable of the aggregate payments.*

Using the normal approximation to Y, calculate the initial size of the fund needed in order to be 95% certain of being able to make the payments for these life annuities.

Answer Let Y_k be the present value random variable of the payment to life k.

$$E[Y_k] = \ddot{a}_x = \frac{1 - A_x}{d} = 11.55$$

$$\text{Var}[Y_k] = \frac{^2A_x - (A_x)^2}{d^2} = \frac{0.22 - 0.45^2}{(0.05 / 1.05)^2} = 7.7175.$$

Then $Y = \sum_{k=1}^{100} Y_k$ is the present value of the aggregate payments.

$$E(Y) = 100\, E(Y_k) = 1155, \text{ and } Var(Y) = 100\, Var(Y_k) = 771.75$$

$$Pr(Y \le F) = Pr\left(Z \le \frac{F - 1155}{\sqrt{771.75}}\right) = 0.95, \text{ thus } \frac{F - 1155}{\sqrt{771.75}} = 1.645.$$

Therefore $F = 1155 + 1.645\sqrt{771.75} = 1200.699$.

Question 11 *For a fully discrete 3-year term insurance of* 1000 *on* (x), *you are given:*

(i) $p_x = 0.975$.
(ii) $i = 0.06$.
(iii) The actuarial present value of the death benefit is 152.85.
(iv) The annual net premium is 56.05.

Calculate P_{x+2}.

Answer

$$\ddot{a}_{x:\overline{3}|} = \frac{\text{Acturial } PV \text{ of the benefit}}{\text{Level annual premium}} = \frac{152.85}{56.05} = 2.727.$$

On the other hand,

$$\ddot{a}_{x:\overline{3}|} = 1 + \frac{0.975}{1.06} + \frac{(0.975)(p_{x+1})}{(1.06)^2} = 2.727.$$

We have $p_{x+1} = 0.93$.

Actuarial PV of the benefit $=$

$$152.85 = 1000 \left[\frac{0.025}{1.06} + \frac{0.975(1-0.93)}{1.06^2} + \frac{(0.975)(0.93)(q_{x+2})}{(1.06)^3} \right].$$

Therefore $q_{x+2} = 0.09$, and $p_{x+2} = 0.91$.

Exam ASTAM topics

The ASTAM Exam is a three-hour exam consisting of 60 points of written-answer questions. The exam's syllabus continues to develop the candidate's knowledge of modeling and important actuarial methods that are useful in modeling, as well as ratemaking and reserving for short-term coverages.

The topics and their weights on the exam follow as they appear on the Spring 2023 syllabus.

1. Topic: Severity Models (12%–22%)
2. Topic: Aggregate Models (12%–22%)
3. Topic: Coverage Modifications (6%–12%)
4. Topic: Construction and Selection of Parametric Models (12%–22%)
5. Topic: Credibility (12%–20%)
6. Topic: Reserving for Short-Term Insurance Coverages (12%–20%)
7. Topic: Pricing for Short-Term Insurance Coverages (4%–12%)

Exam ASTAM sample questions and solutions

Sample questions and solutions demonstrate that written answer questions usually consist of multiple parts to test examinees' skills in several learning objectives from the exam syllabus, and some parts require a nonmathematical narrative to see whether students truly understand the reason or logic behind a certain formula or calculation. The multiple-part structure of a question makes it possible that a student makes an error on the first or second parts of the answer and carries the wrong numerical result into answering the other parts. This is actually a challenge to problem writers since the exam allows students to earn partial credits.

Question 1 *For a portfolio of short-term insurance contracts, claim amounts are modeled with a density function* $f(x; \lambda) = \lambda^2 x e^{-\lambda x}, x > 0, \lambda > 0$

A random sample of n claim amounts, $x_1, x_2, \ldots, x_n$, *is used to estimate the parameter* λ.

(a) *Derive and simplify as far as possible a formula for the maximum likelihood estimate of* λ.

(b) *Use the moment generating function to show that* $Y = 2n\lambda\bar{X}$ *has a Chi-square distribution and states the degrees of freedom.*

(c) *Determine the Fisher information function,* $I(\lambda)$, *and use it to derive a formula for the asymptotic variance of* $\hat{\lambda}$.

(d) *A random sample of* 10 *claim amounts resulted in a sample mean of* $\bar{x} = 358$.

   (i) *Use the results in (b) to calculate an exact* 95% *confidence interval of* λ.

   (ii) *Use the results in (c) to calculate an approximate (asymptotic)* 95% *confidence interval of* λ.

(e) *Comment briefly on the comparison of these confidence intervals.*

Solution

(a) $\ln\big(f(x_i;\lambda)\big) = 2\ln\lambda + \ln x_i - \lambda x_i$

Log-likelihood function

$$l(\lambda) = \sum_{i=1}^{n} \ln\big(f(x_i;\lambda)\big) = 2n \ \text{In} + \sum_{i=1}^{n} \text{In} x_i - \lambda\sum_{i=1}^{n} x_i$$

Take derivative:

$$\frac{dl(\lambda)}{d\lambda} = \frac{2n}{\lambda} - \sum_{i=1}^{n} x_i$$

Set it to zero and solve.

$$\hat{\lambda} = \frac{2n}{\sum_{i=1}^{n} x_i} = \frac{2}{\bar{x}}.$$

(b) $M_Y(t) = E\Big[e^{2n\lambda\bar{X}\,t}\Big] = E\left[e^{2\lambda t\sum_{i=1}^{n}X_i}\right] = \big(E[e^{2\lambda Xt}]\big)^{n}$ since $X_1, X_2,\ldots, X_n$

are independent and identically distributed.

$$E\Big[e^{2\lambda Xt}\Big] = \int_0^{\infty} e^{2\lambda xt}\lambda^2 xe^{-\lambda x}dx = \lambda^2\int_0^{\infty} e^{-\lambda x(1-2t)}xdx$$

$$= \frac{\lambda^2}{\big(\lambda(1-2t)\big)^2}\int_0^{\infty}\big(\lambda(1-2t)\big)^2 xe^{-\lambda x(1-2t)}dx.$$

If $1 - 2t > 0$, the integrand is the pdf of the original distribution, with a new value for parameter, $\lambda(1 - 2t)$, so the integral is equal to 1, leading to

$$E\Big[e^{2\lambda Xt}\Big] = (1 - 2t)^{-2}$$

Hence $M_Y(t) = E\left[e^{2n\lambda \bar{X} t}\right] = (1 - 2t)^{-2n}$, which is the MGF of Chi-square distribution, with degrees of freedom $v = 4n$.

(a) The Information function is $-E\left[\dfrac{d^2 l(\lambda)}{d\lambda^2}\right] = -E\left(\dfrac{2n}{\lambda^2}\right) = \dfrac{2n}{\lambda^2}$

Thus the asymptotic variance of $\hat{\lambda}$ is $\left(-E\left[\dfrac{d^2 l(\lambda)}{d\lambda^2}\right]\right)^{-1} = \dfrac{\lambda^2}{2n}$.

(b) (i) Using the fact that Y follows Chi-square distribution, a 95% confidence interval for Y is $(Q_{0.025}(Y), Q_{0.975}(Y))$, where $Q_\alpha(Y)$ is the α-percentile of Chi-square distribution with $v = 4n$ degrees of freedom.

Using CHISQ.INV in Excel (ASTAM is computer-based), we find the 95% CI for $Y = 2n\lambda \bar{X}$ is (24.43, 59.34). Divide it by $2n\bar{x}$ gives a 95% confidence interval for λ of (0.00341, 0.00829).

(ii) The MLE is asymptotic and normally distributed. We have $\hat{\lambda} = \dfrac{2}{\bar{x}} = \dfrac{2}{358} = 0.005587$. The variance is approximately

$\dfrac{\lambda^2}{2n} = 1.5605 * 10^{-6}$, making the standard deviation 0.00125. So the

approximate 95% CI is $0.005587 \pm 1.96 * 0.00125 = (0.00314, 0.00804)$.

(c) The confidence intervals are quite close. The MLE CI is an asymptotic result, and we would expect less accuracy applying this to a sample of only 10 values. The chi-square confidence interval is an exact interval for Y, based on the actual sample size rather than on asymptotic results, so we would expect it to be more accurate.

Question 2 *Santiago Substandard Auto Company (SSAC) sells automobile insurance policies to drivers who have a bad driving record. Each policy covers only one automobile.*

During 2021, SSAC had 10,000 policies in force. Each policy covered 100% of all losses, with no deductibles, no upper limit, and no coinsurance.

During 2021, the distribution of the number of claims per insured automobile for SSAC was:

Number of claims in 2021	Number of policies
0	2400
1	3400
2	2400
3	1500
4	300

(a) *Assume that the claim frequency is distributed as a Poisson random variable with parameter λ. Calculate the 90% linear confidence interval for λ based on the previous data.*

(b) *SSAC wishes to use the data to test the following hypothesis, using the Chi-square test, with a 99% significance level:*

   H_0: The data is from a Poisson distribution.

   H_1: The data is not from a Poisson distribution.

   The data for 3 and 4 claims in 2021 are combined for this test.

   (i) *Calculate the Chi-square test statistic.*

   (ii) *Calculate the p-value for this test and state your conclusion with regard to the hypothesis.*

(c) *You are given the following claim severity sample from 2021:*

$$200 \quad 1000 \quad 5000 \quad 10{,}000 \quad 100{,}000$$

SSAC uses this data to test the following hypothesis, using the Kolmogorov Smirnov Test, with a 95% significance level:

H_0: The data is from a Pareto distribution with parameters $\theta = 75{,}000$ and $\alpha = 4$.

H_1: The data is not from a Pareto distribution with parameters $\theta = 75{,}000$ and $\alpha = 4$.

(i) *Calculate the Kolmogorov-Smirnov test statistic.*

(ii) *You are given that the 5% critical value for the Kolmogorov-Smirnov test is approximately $1.36/\sqrt{n}$. State your conclusion.*

(d) *In selecting a model, there are two important concepts. The first concept is known as parsimony while the second does not have a specific name.*

   (i) *State the principle of parsimony.*

   (ii) *State the second concept.*

   (iii) *Identify the score-based approach that is consistent with parsimony and state why it is consistent.*

Solution

(a) $\hat{\lambda} = \bar{x} = \dfrac{3400 + 2(2400) + 3(1500) + 4(300)}{10{,}000} = 1.39$

$$\mathrm{Var}\left(\hat{\lambda}\right) = \frac{\hat{\lambda}}{n} = \frac{1.39}{10{,}000} = 0.000139$$

Thus 90% confidence interval is

$1.39 \pm 1.645\sqrt{0.000139} = (1.37061, 1.40904).$

(b) Using $\hat{\lambda} = 1.39$ from (a):

Number of claims	Number of policies observed	Number of policies expected	$\dfrac{\left(E_j - O_j\right)^2}{E_j}$
0	2400	$10,000e^{-1.39} = 2490.753$	3.307
1	3400	$10,000e^{-1.39}(1.39) = 3462.147$	1.116
2	2400	$10,000e^{-1.39}\left(\dfrac{1.390^2}{2}\right) = 2406.192$	0.016
3+	1800	$10,000 - 2490.7533462.147 - 2406.192$ $= 1640.908$	15.425
		Test statistics =	19.864

Degrees of freedom = $4 - 1 - 1 = 2$. The P-value is $\Pr\left[\chi_2^2 > 19.864\right] =$ $= 5 * 10^{-5}$. Since it is very small, we reject the null hypothesis. It is extremely unlikely that this data came from the Poisson distribution with mean 1.39.

(c) The K-S test statistic calculations follow.

x	$F_n(x-)$	$F_n(x)$	$F^*(x) = 1 - \left(\dfrac{75,000}{75,000 + x}\right)^4$	Absolute value of maximum difference
200	0	0.2	$1 - \left(\dfrac{75,000}{75,200}\right)^4 = 0.01060$	0.18940
1000	0.2	0.4	$1 - \left(\dfrac{75,000}{76000}\right)^4 = 0.05160$	0.34840
5000	0.4	0.6	$1 - \left(\dfrac{75,000}{80000}\right)^4 = 0.22752$	0.37248
10,000	0.6	0.8	$1 - \left(\dfrac{75,000}{85,000}\right)^4 = 0.39387$	0.40613
100,000	0.8	1.0	$1 - \left(\dfrac{75,000}{175,000}\right)^4 = 0.96626$	0.16626

The test statistic D is the maximum absolute difference, i.e., $D = 0.40613$.

The critical value is $\dfrac{1.36}{\sqrt{5}} = 0.6082$. Since $D < 0.6082$ we do not reject H_0.

(i) Unless there is considerable evidence to do otherwise, the simpler model is preferred.

(ii) If you try enough models, one will look good, even if it is not.

(iii) Choose the model with the highest p-value under the chi-square goodness of fit test. This is consistent with parsimony because the more complex tests have less degrees of freedom.

Question 3

(a) *There are four essential objectives of Ratemaking. Describe each one and explain why these objectives are essential.*

(b) *You are the rate-making actuary for ABC Auto. You are setting rates for the auto coverage, which is a short-term insurance product. You are given the following data:*

Calendar year	Earned premium
2017	10,000
2018	12,000
2019	8000

Assume that all policies are one-year policies and that policies are issued uniformly throughout the year.

The following rate changes have occurred:

Date	Rate change
March 15, 2017	10% increase
July 30, 2018	8% increase
October 19, 2019	4% decrease

(i) *Graphically display the parallelogram that would be used to apply the parallelogram method.*

(ii) *Using the parallelogram method, calculate the earned premium for 2017, 2018, and 2019 based on current rates.*

(c) *For collision coverage for ABC Auto, rate differentials for 2019 are developed using the following information:*

Type	Proportion of total drivers	Claim frequency Poisson annual	Severity Gamma
Safe	0.5	0.04	$\alpha = 3 \ \theta = 1000$
Not So Safe	0.3	0.10	$\alpha = 4 \ \theta = 1000$
Reckless	0.2	0.25	$\alpha = 5 \ \theta = 1000$

(i) Calculate the differentials for each type, given that Safe is the base rate.

(ii) During 2019, ABC Auto experienced the following loss ratios for collision coverage based on the indicated differentials developed in Question 16:

Type	Loss ratio
Safe	60%
Not So Safe	63%
Reckless	51%

Use the loss ratio method to determine the new indicated differentials for 2020 for the Not So Safe and Reckless categories.

Solution

(a) The following are the four essential objectives of ratemaking.

1. The rates must cover the expected losses and expenses; for the insurer to stay in business, income (premiums and investment income) must at least equal outgo (losses and expenses).

2. Ratemaking should produce rates that make adequate provision for contingencies. While the rates should cover the expected losses, rates should also build in the cost of the unexpected (e.g., unusual weather patterns, wildfires, 100-year floods). Such provisions must be established while maintaining competitive rates without endangering solvency.

3. Rates should encourage loss control. Rates should lead to a risk categorization that provides policyholders strong economic incentives to help reduce claim frequency and/or severity. Such process not only allows for the insurer to lower rates but also extends service to society by encouraging good behavior that reduces accidents, injuries, and property damage.

4. Rates must satisfy rate regulators. Almost all rates must be filed with and approved by the state insurance department or other agencies. Any proposed rate changes must come with supporting full actuarial documentation. The basic requirement of rate regulation is that rates must be adequate, not excessive, and not unfairly discriminatory.

(b) See the diagram below.

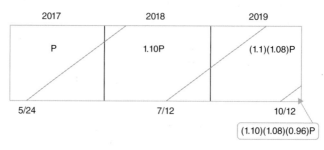

For 2017:

$$\text{Weighted premium} = \left[\frac{\left(\frac{19}{24}\right)\left(\frac{19}{24}\right)}{2}\right]1.10P + \left[1 - \frac{\left(\frac{19}{24}\right)\left(\frac{19}{24}\right)}{2}\right]P = 1.031337P$$

$$\text{Current rate earned premium} = (10,000)\left[\frac{(1.10)(1.08)(1-0.04)}{1.031337}\right] = 11,058.27$$

For 2018:

$$\text{Weighted premium} = \left[\frac{\left(\frac{5}{24}\right)\left(\frac{5}{24}\right)}{2}\right]P + \left[\frac{\left(\frac{5}{12}\right)\left(\frac{5}{12}\right)}{2}\right](1.10)(1.08)P$$

$$+ \left[1 - \frac{\left(\frac{5}{24}\right)\left(\frac{5}{24}\right)}{2} - \frac{\left(\frac{5}{12}\right)\left(\frac{5}{12}\right)}{2}\right](1.10)P = 1.105469P$$

$$\text{Current rate earned premium} = (12,000)\left[\frac{(1.10)(1.08)(0.96)}{1.105469}\right] = 12,380.05$$

For 2019:

$$\text{Weighted premium} = \left[\frac{\left(\frac{7}{12}\right)\left(\frac{7}{12}\right)}{2}\right](1.10)P + \left[\frac{\left(\frac{2}{12}\right)\left(\frac{2}{12}\right)}{2}\right](1.10)(1.08)(0.96)P$$

$$+ \left[1 - \frac{\left(\frac{7}{12}\right)\left(\frac{7}{12}\right)}{2} - \frac{\left(\frac{2}{12}\right)\left(\frac{2}{12}\right)}{2}\right](1.10)(1.08)P = 1.172368P$$

$$\text{Current rate earned premium} = (8,000)\left[\frac{(1.10)(1.08)(0.96)}{1.172368}\right] = 7,782.40$$

(c) For each type, average severity $E(Y) = \alpha\theta$ using the formula for Gamma distribution.

We need to calculate the expected loss for each type using $E(S) = E(N)E(Y)$.

Safe: $(0.04)(3000) = 120$

Not So Safe: $(0.10)(4000) = 400$

Reckless: $(0.25)(5000) = 1250$

(i) Differential for Safe type is 1.00 since it is the base type.

Differential for Not So Safe type is $\dfrac{400}{120} = 3.3333$.

Differential for Reckless type is $\dfrac{1250}{120} = 10.4167$.

(ii) Indicated Differential for Not So Safe type is $(3.3333)\dfrac{0.63}{0.60} = 3.5000$.

Indicated Differential for Reckless type is $(10.4167)\dfrac{0.51}{0.60} = 8.5242$.

Question 4 *During 2021, AAA Insurance collects the following premium amounts:*

Month	January	March	May	July
Premium collected	1600	1800	1800	1200

You are given:

(i) All premiums are paid on the first day of the month and all premiums are annual premiums.

(ii) The unearned premium reserve on December 31, 2020 was 1300.

(iii) The company expects a loss ratio of 80%.

(iv) During 2021, the company paid losses for claims incurred in 2021 of 2500.

  (a) Use the loss ratio method to determine the outstanding claim reserve for accident year 2021 for this coverage that should be held on December 31, 2021.

  (b) The insurer has the following paid claims triangle for this coverage. There is no further development after year 5.

Cumulative loss payments by development year						
	Development year					
Accident year	**0**	**1**	**2**	**3**	**4**	**5**
2016	1000	1500	1700	1800	1850	1875
2017	1100	1750	1775	1825	1870	
2018	1200	1900	2200	2350		
2019	1500	2200	2500			
2020	2000	2900				
2021	2500					

(i)   *The loss development factor for the period from year 0 to year 1 is 1.51 to two decimal places. Calculate the loss development factor to five decimal places.*
(ii) *Calculate the outstanding claim reserve on December 31, 2021.*
(iii) *Determine the total amount of claims paid in 2021.*
    (a) *For the claims from accident year 2021, determine the reserves as of December 31, 2021 using the Bornhuetter-Ferguson method, where the prior estimate is the outstanding claim reserve using the loss ratio method.*
    (b) *Explain briefly one advantage and one disadvantage of the Bornhuetter-Ferguson method, compared with the Chain Ladder approach.*

Solution

(a) The earned part of the premium paid in January is 1600.

The earned part of the premium paid in March is $\dfrac{10}{12} * 1800 = 1500.$

The earned part of the premium paid in May is $\dfrac{8}{12} * 1200 = 800.$

The earned part of the premium paid in July is $\dfrac{6}{12} * 1200 = 600$.

Thus the total earned premium paid in 2021 is 4500, adding 1300, the premium paid in 2020 and earned in 2021, the total earned premium in 2021 is 5800.

Because expected total losses are 5800*0.8 = 4640, therefore

Reserve = Expected total losses Claims − Already paid = 4640 − 2500 = 2140.

(b) The loss development factor for the periods follow.

  (i) The loss development factor for the period from year 0 to year 1:

$$f_1 = \frac{1500 + 1750 + 1900 + 2200 + 2900}{1000 + 1100 + 1200 + 1500 + 2000} = 1.50735.$$

From year 1 to year 2:

$$f_2 = \frac{1700 + 1775 + 2200 + 2500}{1500 + 1750 + 1900 + 2200} = 1.11224.$$

From year 2 to year 3:

$$f_3 = \frac{1800 + 1825 + 2350}{1700 + 1775 + 2200} = 1.05286.$$

From year 3 to year 4:

$$f_4 = \frac{1850 + 1870}{1800 + 1825} = 1.02621.$$

From year 4 to year 5:

$$f_5 = \frac{1875}{1850} = 1.01351.$$

(ii) The computations are displayed in the table below.

| Accident year | Development year | | | | | | Reserve |
	0	1	2	3	4	5	
2016	1000	1500	1700	1800	1850	1875	
2017	1100	1750	1775	1825	1870	1895.26	25.26
2018	1200	1900	2200	2350		2444.15	94.15
2019	1500	2200	2500			2737.60	237.60
2020	2000	2900				3532.05	630.15
2021	2500					4589.73	2089.73

For example, the entries for 2021, 4589.69 and 2089.73 are computed as $2{,}500\, f_1 f_2 f_3 f_4 f_5$ and $2089.73 = 4589.73 - 2500$.

For 2017, the value 1895.26 is obtained as 1870 (f_5), and the reserve 25.26 is the result of subtracting 1870 from 1895.26.

The outstanding claim reserve on December 31, 2021 is the sum of reserves for years 2017 through 2021, equal to 3076.89.

(i) The total sum paid in 2021 is:

$$2500 + (2900 - 2000) + (2500 - 2200) + (2350 - 2200)$$
$$+ (1870 - 1825) + (1875 - 1850) = 3920.$$

(a) The reserve by loss ratio method is $5800 * 0.8 = 4640$.

The ultimate loss development factor for 2021 is $f_{ult} = f_1 f_2 f_3 f_4 f_5 = 1.835892$.

BF estimate of projected 2021 AY claims is $(4640) \times \left(1 - \dfrac{1}{f_{ult}}\right) = 2112.62$.

(b) BF Advantage – much less reliance on a single data point in the most recent AY.

BF Disadvantage – relies on subjective estimate of loss ratio (and adequacy of premiums).

Question 5 *You are given that $X_1, X_2, \ldots, X_n$ are independent and identically distributed copies of a continuous loss random variable X, where X represents a firm's aggregate daily losses from operational risks. Let $M_n = \max(X_1, X_2, \ldots, X_n)$.*

The n-day Expected Maximum (nDEM) risk measure is defined as $\rho(X) = E(M_n)$. *Currently, the firm uses the 95% VaR risk measure, denoted* $Q_{0.95}$ *A colleague suggests changing to the nDEM risk measure, using* $n = 20$ *for consistency with the current 1-in-20-day standard.*

(a) Show that $\Pr(M_n > Q_{0.95}) = 0.64$ to the nearest 0.01. You should calculate the probability to the nearest 0.001.

Now suppose that X is exponentially distributed with mean 1. Let
$$\alpha(n) = \frac{n-1}{n}.$$

(b) Show that the VaR of X at $\alpha(n)$ is $Q_{\alpha(n)} = \log(n)$.

(c) Show that the $\alpha(n)$-Expected Shortfall of XX is $ES_{\alpha(n)} = 1 + \log(n)$. You may use without proof the memoryless property of the exponential distribution.

(d) You are given that
- The exponential distribution is in the Maximum Domain of Attraction (MDA) of the Gumbel distribution.
- The mean of the standard Gumbel distribution is 0.5772.
- The normalizing functions for the exp(1) distribution under the Fisher-Tippett-Gnedenko theorem, are $c_n = 1$ and $d_n = \log n$.
  - (i) Explain what it means to say that the exponential distribution is in the MDA of the Gumbel distribution.
  - (ii) Assume that n is sufficiently large that the normalized n-day maximum loss can be approximated by the standard Gumbel distribution. Show that
$$Q_{\alpha(n)}(X) < E[M_n] < ES_{\alpha(n)}(X).$$

(e) Explain why the ordering in (d) is likely to hold more generally.

Solution

(a) $\Pr[M_{20} > Q_{0.95}(X)] = 1 - \Pr[M_{20} \le Q_{0.95}(X)] = 1 - [F_X(Q_{0.95}(X))]^{20}$
$$= 1 - (0.95)^{20} = 0.64151$$

(b) $\alpha(n) = \dfrac{n-1}{n} = \Pr[X \le Q_{\alpha(n)}(X)] = 1 - e^{-Q_{\alpha(n)}(X)}$

Thus $\dfrac{1}{n} = e^{-Q_{\alpha(n)}(X)}$, or $e^{Q_{\alpha(n)}(X)} = n$, therefore $\ln n = Q_{\alpha(n)}(X)$.

(c) Since it is continuous,
$$ES_{\alpha(n)}(X) = E\left[(X)|X > Q_{\alpha(n)}(X)\right]$$
$$= Q_{\alpha(n)}(X) + E\left[X - Q_{\alpha(n)}(X)|X > Q_{\alpha(n)}(X)\right]$$

$$= Q_{\alpha(n)}(X) + E[X] = \ln n + 1,$$ where the second from last equality is based on the memoryless property of exponential distribution.

(d) The exponential distribution is in the MDA of the Gumbel EV distribution means that there exist deterministic functions c_n and d_n, such that as $n \to \infty$, the distribution of $\dfrac{M_n - d_n}{c_n}$ converges to the standard Gumbel distribution.

Let $H(x)$ denote the density function of the Gumbel distribution.

$$Pr\left[\frac{M_n - d_n}{c_n} \leq x\right] \approx H(x)$$

Hence $E\left[\dfrac{M_n - d_n}{c_n}\right] \approx 0.5772$, or $E[M_n - \ln n] \approx 0.5772$, thus

$$E[M_n] \approx 0.5772 + \ln n.$$

Therefore we have

$$\left[Q_{\alpha(n)}(X) = \ln n\right] < \left[E[M_n] = 0.5772 + \ln n\right] < [ES_{\alpha(n)}(X) = 1 + \ln n].$$

(e) Suppose we have a sample of $20N$ values of X_i, split into N blocks of 20, where N is a large number.

We would approximate $E[M_{20}]$ by taking the average of the N block maxima. As $N \to \infty$, this approximation will converge to the true value.

We would approximate $ES_{0.95}$ by taking the average of the largest N values. As $N \to \infty$, this approximation will converge to the true value. This cannot be smaller than the $E[M_{20}]$ estimate (as the boundary case is that the N largest values are also block maxima) but it could be bigger, if one or more blocks have several values that are larger than the smallest block maximum. Hence, $ES_{0.95}$ must be greater than $E[M_{20}]$, unless both are equal to the maximum possible value of X.

Also, for each block, the estimated 95% quantile of X lies between the 19th and 20th values (the smoothed empirical estimate would be the 19.95th value, estimated by linear interpolation). That means that the expected value of the 95% quantile is less than or equal to the expected value of the block maximum, with a block size of 20, i.e., that $E[M_{20}]$ must be greater than $Q_{0.95}$, unless both are equal to the maximum possible value of X.

Exam ALTAM topics

Similar to Exam ASTAM, the ALTAM Exam is a three-hour exam consisting of 60 points of written-answer questions. The syllabus for Exam ALTAM continues to develop the candidate's knowledge of the theoretical basis of contingent payment models and the application of those models to insurance and other financial risks. There are seven topic areas on the spring 2023 syllabus, and several learning objectives for each topic area.

1. Topic: Survival Models for Multiple State Contingent Cashflows (10%–20%)
2. Topic: Premium and Policy Valuation for Long-Term State-Dependent Coverages (12%–20%)
3. Topic: Joint Life Insurance and Annuities (8%–16%)
4. Topic: Profit Analysis (10%–20%)
5. Topic: Pension Plans and Retirement Benefits (10%–18%)
6. Topic: Universal Life Insurance (10%–18%)
7. Topic: Embedded Options in Life Insurance and Annuity Products (10%–18%)

Exam ALTAM sample questions and solutions

Question 1 *For a special fully discrete two-year term insurance on an individual, age 60, you are given:*

(i) *The death benefit is 1000 plus the return of gross premiums paid with interest at 6%.*

(ii) *The following double decrement table, where decrement (d) is death and decrement (w) is withdrawal:*

x	$q_x^{(d)}$	$q_x^{(w)}$
60	0.06	0.04
61	0.12	0.00

(iii) *There are no withdrawal benefits.*
(iv) *$i = 0.06$*
(v) *G denotes the annual gross premium.*
(vi) *L_0 denotes the insurer's loss at issue random variable for the individual's policy.*

(a) *Calculate the values in the following table. Express L_0 in terms of G where appropriate:*

Event	Value of L_0 given that the event occurred	Probability of the event
Death in year 1		
Withdrawal in year 1		
Death in year 2		
Neither death nor withdrawal		

(b) There are two parts.

> *(i) Show that $E(L_0) = a - bG$, where a is 150 to the nearest 10 and b is 1.58 to the nearest 0.01. You should calculate a to the nearest 1 and b to the nearest 0.001.*

> *(ii) Show that $\text{Var}(L_0) = cG^2 + dG + e$, where c is 0.5 to the nearest 0.1, d is 480 to the nearest 10, and e is 116,000 to the nearest 1000. You should calculate c to the nearest 0.01, d to the nearest 1, and e to the nearest 100.*

(c) The insurer expects to issue 200 such policies to insureds with independent future lifetimes. The premium for each policy is $G = 130$.

Let L_{agg} denotes the insurer's aggregate future loss random variable at issue for these 200 policies. Calculate $\Pr(L_{\text{agg}} > 0)$ using the normal approximation without continuity correction.

Solution

(a) Noting $v = 1.06^{-1} = 0.9434$,

Death in year 1:

$$L_0 | \text{Event} = (1000 + G(1 + i))v - G = 1000v = 943.4.$$

Probability $= 0.06$.
Withdrawal in year 1:

$$L_0 | \text{Event} = -G$$

Probability $= 0.04$.
Death in year 2:

$$L_0 | \text{Event} = \left(1000 + G(1 + i) + G(1 + i)^2\right)v^2 - G(1 + v) = 1000v^2 = 890.0$$

Probability $= (0.90)(0.12) = 0.108$.
Survival in force to the end of year 2:

$$L_0 | \text{Event} = -G(1 + v) = -1.9434G$$

Probability $= (0.9)(1 - 0.12) = 0.792$.

Fill in the table:

Event	Value of L_0 given that the event occurred	Probability of the event
Death in year 1	943.4	0.06
Withdrawal in year 1	$-G$	0.04
Death in year 2	890.0	0.108
Neither death nor withdrawal	$-1.9434G$	0.792

(b) The solution follows.

(i) $E(L_0) = (943.4)(0.06) + (-G)(0.04) + (890.0)(0.108)$
$$+ (-1.9434G)(0.792) = 152.7 - 1.579G.$$

Thus $a = 152.7,\ b = 1.579$

(ii) $E(L_0^2) = (943.4)^2 (0.06) + (-G)^2 (0.04) + (890.0)^2 (0.108)$
$$+ (-1.9434G)^2 (0.792) = 138,947 + 3.0312G^2$$

$\text{Var}(L_0) = E(L_0^2) - [E(L_0)]^2 = 115,630 + 482.G + 0.538G^2.$

Thus $c = 0.538,\ d = 482.2,\ e = 115,630$

(c) For each policy:

$G = 130$

$E(L_0) = 152.7 - 1.579G = -52.57$

$\text{Var}(L_0) = E(L_0^2) - [E(L_0)]^2 = 115,630 + 482.G + 0.538G^2 = 187,408.$

So for the aggregate loss:

$E(L_{\text{agg}}) = (200)(-52.57) = -10,514$

$\text{Var}(L_{\text{agg}}) = (200)(187,408) = 37,481,600 = (6122.2)^2.$

Using normal approximation,

$$\Pr(L_{\text{agg}} > 0) = 1 - \Phi\left(\frac{0 - (-10,514)}{6122.2}\right) = 1 - \Phi(1.72)$$

$$= 1 - 0.9573 = 0.0427.$$

Question 2 For (x) and (y) with independent future lifetimes, you are given that $q_x = 0.2$ and $q_y = 0.1$.

(a) Explain in words the meaning of the probability described by the symbol q_{xy}.
You are also given that mortality within integral ages follows a uniform distribution of deaths assumption for each of (x) and (y) individually.

(b) Sketch the graph of $_s p_x$ as a function of s for $0 \le s \le 1$. You should mark numerical values on each axis.

(c) Show that $_s q_{xy} = s\, q_{xy} + g(s) q_{\overline{xy}}$ for $0 \le s \le 1$, where for $g(s)$ is a function of s that you should specify.

Solution

(a) The symbol q_{xy} is the probability that at least one of the two lives, currently age x and y, dies within one year.

(b) The diagram follows where the horizontal axis is s (years) and the vertical is probability.

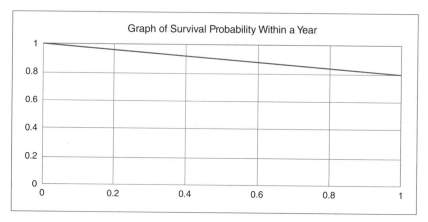

(c) We have

$$_s q_{xy} = 1 - {}_s p_x \, {}_s p_y \text{ (independence)}$$

$$= 1 - \left(1 - {}_s q_x\right)\left(1 - {}_s q_y\right)$$

$$1 - \left(1 - s^* q_x\right)\left(1 - s^* q_y\right) (UDD)$$

$$= s\left(q_x + q_y\right) - s^2 q_x q_y$$

Also $q_x + q_y = q_{\overline{xy}} + q_{\overline{xy}}$ and $q_{\overline{xy}} = q_x q_y$,

Hence $_s q_{xy} = s^* q_{xy} + s\, q_{\overline{xy}} - s^2 q_{\overline{xy}}$, therefore $g(s) = s - s^2$.

Question 3 *You are using the following 3-state Markov model to price a 10-year disability insurance product.*

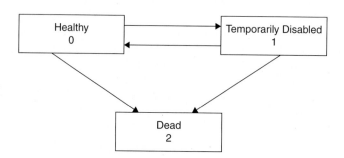

(a) *Show that for this model,* $\displaystyle\sum_{j=0}^{2} \bar{a}^{0j}_{x:\overline{10|}} = \bar{a}_{\overline{10|}}$.

The product has the following features:
- *The product is issued to individuals age x who are in the Healthy state.*
- *The product pays a continuous disability benefit at a rate of 1000 per year while the insured is in the Temporarily Disabled state.*
- *The product pays a death benefit of 10,000 at the moment of death.*
- *Net premiums are payable continuously while the insured is in the Healthy state.*

You are also given the following information:

$$\delta = 0.1 \qquad \bar{a}^{00}_{x:\overline{10|}} = 4.49 \quad \bar{a}^{02}_{x:\overline{10|}} = 1.36 \quad \bar{A}^{02}_{x:\overline{10|}} = 0.3871$$

$$\mu^{01}_{x+t} = 0.04 \quad \mu^{02}_{x+t} = 0.02t \quad \mu^{10}_{x+t} = 0.05 \quad \mu^{12}_{x+t} = 0.04t$$

(a) *Show that the net premium rate for this policy is 970 per year to the nearest 10. You should calculate the rate to the nearest 1.*

Let $_tV^{(i)}$ *denote the net premium policy value for a policy in state* i *at time t. You are given:*

$$_3V^{(0)} = 1304.54 \quad _3V^{(1)} = 7530.09$$

(b) *Calculate* $\dfrac{d}{dt}\,_tV^{(0)}$ *at* $t = 3$.

(c) *Your company is considering adding an additional feature to this product. Under this additional feature, the insurer would return the sum of the premiums paid at the end of 10 years without interest if no benefits were paid during the life of the policy.*

Calculate the increase in the net premium rate payable continuously for the product as a result of including this feature.

Solution

(a) $\displaystyle \bar{a}^{0j}_{x:\overline{10|}} = \int_0^{10} {}_tp^{0j}_x\, e^{-\delta t}\, dt$

So $\displaystyle \bar{a}^{00}_{x:\overline{10|}} + \bar{a}^{01}_{x:\overline{10|}} + \bar{a}^{02}_{x:\overline{10|}} = \int_0^{10} [{}_tp^{00}_x + {}_tp^{01}_x + {}_tp^{02}_x]e^{-\delta t}\, dt$

But $\displaystyle {}_tp^{00}_x + {}_tp^{01}_x + {}_tp^{02}_x = 1$

Hence $\displaystyle \bar{a}^{00}_{x:\overline{10|}} + \bar{a}^{01}_{x:\overline{10|}} + \bar{a}^{02}_{x:\overline{10|}} = \int_0^{10} e^{-\delta t}\, dt = \bar{a}_{\overline{10|}}$.

(b) EPV Premiums: $P\bar{a}^{00}_{x:\overline{10}|} = 4.49P$

$$\bar{a}_{\overline{10}|} = \frac{1 - v^{10}}{\delta} = 6.32121$$

EPV Disability Annuity: $1000\bar{a}^{01}_{x:\overline{10}|} = 1000\left[\bar{a}_{\overline{10}|} - \bar{a}^{00}_{x:\overline{10}|} - \bar{a}^{02}_{x:\overline{10}|}\right]$

$$= 1000\left[6.32121 - 4.49 - 1.36\right] = 471.21$$

EPV Death benefit: $10,000\bar{A}^{02}_{x:\overline{10}|} = 3871.0$

Hence the premium is $P = \dfrac{3871 + 471.21}{4.49} = 967.1$

(c) The Thiele equation at time t is:

$$\frac{d}{dt}\,_tV^{(0)} = \delta\,_tV^{(0)} + P - \mu^{01}_{x+t}\left(\,_tV^{(1)} - \,_tV^{(0)}\right) - \mu^{02}_{x+t}(10,000 - \,_tV^{(0)})$$

At $t = 3$,

$$\frac{d}{dt}\,_tV^{(0)}\bigg|_{t=3} = (0.1)(1304.54) + 967.1$$

$$-(0.04)(7530.09 - 1304.54) - 3(0.2)(10,000 - 1304.54) = 326.80.$$

(d) Let P^* denotes the new premium. The EPV of the return of premium benefit is:

$$10P^*e^{-10\delta}\,_{10}p^{\overline{00}}_x, \quad \text{where}$$

$$_{10}p^{\overline{00}}_x = \exp\left(-\int_0^{10}\left[\mu^{01}_{x+t} + \mu^{02}_{x+t}\right]dt\right) = \exp\left(-\int_0^{10}[0.04 + 0.02t]dt\right)$$

$$= e^{-1.4} = 0.24660$$

$$EPV = 10P^*e^{-10(0.1)}(0.24660) = 0.90718P^*$$

EPV of new premium: $P^*\bar{a}^{00}_{x:\overline{10}|} = 4.49P^*$

EPV of new benefit = EPV of original benefit + EPV of return premium benefit, therefore

$$P^* = \frac{3871 + 471.21}{4.49 - 0.90718} = 1212.0$$

The increase in the net premium is $P^* - P = 244.9.$

Question 4 *For a special 3-year term life insurance issued to (50) with a premium refund feature, you are given:*

(i)   The death benefit is 100,000.

(ii) The premium refund feature refunds the last premium payment, without interest, at the end of the 3-year term if the insured is still alive.

(iii) The mortality rates are:

x	q_x
50	0.00592
51	0.00642
52	0.00697

(i)   Precontract expenses are 155.

(ii) Commissions are 5% of each premium.

(iii) The hurdle rate is 14%.

(iv) The reserves of this policy have been set to:

t	$_tV$
0	0
1	400
2	800

(i)   The annual premium for this policy is 1100.

(ii) The earned interest rates are:

Year 1	Year 2	Year 3
0.01	0.02	0.03

(a) Show that the expected profit in policy year 2 for a policy in force at the start of year 2 is 37 to the nearest 1. You should calculate your answer to the nearest 0.01.

(b) Calculate the profit vector of this policy.

(c) Calculate the profit signature and Net Present Value (NPV) of this policy.

(d) Rank from low to high the Internal Rate of Return (IRR) of the following products, explaining your order.

Product A: The special 3-year term life insurance described above.

Product B: A 3-year term life insurance policy with the profit signature: [−155, 0, 0, 210].

Product C: The same special 3-year term life insurance as Product A, except that the reserves of the product have been set to:

t	$_tV$
0	0
1	300
2	800

Solution

(a)

$$Pr_2 = \left(_1V + P - E \right)\left(1 + i_2\right) - q_{51}\ S - p_{51}\left(_2V \right)$$
$$= \left(400 + 1100 - 55\right)\left(1.02\right) - \left(0.00642\right)\left(100,000\right) - \left(1 - 0.00642\right)\left(800\right)$$
$$= 374.$$

(b) In the table below P is the premium

E_t denotes expenses.

I_t denotes interest on fund in year t .

Expected death benefit $EDB_t = 100,000 q_{50+t}$.

In year 3, Expected return premium benefit $EMB_t = 1,000 p_{52}$.

$$E\left(_tV \right) = p_{50+t-1}\ _tV$$

The profit vector is the final column.

t	$_{t-1}V$	P	E_t	I_t	EDB_t	EMB_t	$E\left(_tV \right)$	Pr_t
0			155					−155.00
1	0	1100	55	10.45	592		397.63	65.82
2	400	1100	55	28.90	642		794.86	37.04
3	800	1100	55	55.35	697	1092.33	0.00	111.02

(c) The profit signature at t is Π_t where:

$$\Pi_0 = Pr_0 \text{ and } \Pi_t = {}_{t-1}p_{50}Pr_t$$

Hence $\left(\Pi_0, \Pi_1, \Pi_2, \Pi_3 \right) = (-155.00, 65.82, 36.82, 109.65)$

And $NPV = \sum_{k=0}^{3} {}_k v_{0.14}^k$

$= -155 + 65.82(1.14)^{-1} + 36.82(1.14)^{-2} + 109.65(1.14)^{-3} = 5.08.$

(d) The IRR of B is the root j to the equation $155 = 210(1 + j) - 3$, and the value is $j = 10.65\%$.

The IRR of A is greater than 14%, because the NPV at 14% is positive. Hence the IRR of B is less than the IRR of A.

Product C has lower reserves in year 1, which allows an earlier release of surplus compared to Product A, which gives a higher NPV than A at a 14% hurdle rate, but does not necessarily mean that C has a higher IRR. The lower reserve in year 1 results in the following profit signature for C: $(-155.00, 165.23, -64.58, 109.65)$.

We note that the NPV of A is a little larger than 14%, because the NPV at 14% is close to zero. Calculating the NPV for A and C at 16% gives -0.65 for A and 9.7 for C. Hence, the IRR for C is greater than 16%, and for A is less than 16%.

Therefore the $IRR(B) < IRR(A) < IRR(C)$.

Question 5 *A benefit plan provides a retirement benefit if the employee lives to age 65, and a death benefit if the employee dies prior to age 65.*

- *The retirement benefit is an annual whole life annuity-due of 3% of the final 3-year average salary for each year of service.*
- *The death benefit is a lump sum payable at the end of the year of death equal to two times the employee's annual salary in the year of death.*

You are given:

- A person started the employment on January 1, 1990 at exact age 38 with a starting salary of 50,000.
- The company gives salary increases of 3% on January 1 each year.
- Employees can terminate employment only by retirement at 65 or death.

You are given the following:

- $q_{62+k} = 0.08 + 0.01k$ for $k = 0,1,2,3$
- The expected present value on January 1, 2017 of an annuity due of 1 per year payable annually to the person, if they survive, will be 4.7491.
- $i = 0.04$

   (a) Calculate the actuarial present value on January 1, 2014 of the person's death benefit.

   (b) Calculate the actuarial present value on January 1, 2014 of the person's retirement benefit.

Solution

(a) Let S_x denote the salary earned in year of age x to x + 1. We have

$$S_x = 50,000(1.03)^{x-38}$$

The EPV of the death benefit is:

$$2S_{62}\, q_{62} \left(1.04\right)^{-1} + \left(2S_{63}\right)\,{}_1|\,q_{62}\left(1.04\right)^{-2} + \left(2S_{64}\right)\,{}_2|\,q_{62}\left(1.04\right)^{-3}$$

where
$$q_{62} = 0.08,\ {}_1|\,q_{62} = (0.92)(0.09) = 0.0828,\ {}_2|\,q_{62} = (0.920(0.91)(0.10)$$
$$= 0.08372$$

So the EPV of the death benefit is

$$2(50,000)(1.03)^{24}\Big[(0.08)(1.04)^{-1} + (1.03)(0.0828)(1.04)^{-2}$$
$$+(1.03)^2(0.08372)(1.04)^{-3}\Big] = 47,716.$$

(b) The EPV of the retirement benefit is:

$$(0.03)(27)(FAS)\big({}_3p_{62}\big)(1.04)^{-3}\,\ddot{a}_{65}$$

Where *FAS* is the final average salary:

$$FAS = 50,000\left(\frac{(1.03)^{24} + (1.03)^{25} + (1.03)^{26}}{3}\right) = 104,719.$$

So the EPV is

$$(0.03)(27)(104,719)(0.75348)(0.8890)(4.7491) = 269,833.$$

2.15   Modern Actuarial Statistics

The Casualty Actuarial Society requires candidates for associateship to have strong foundations in statistics as the knowledge and skills are extensively applied to modeling of property and liability losses and pricing of insurance products covering such losses. A candidate's aptitude in statistics is tested on exams Modern Actuarial Statistics (MAS) I and II.

Modern actuarial statistics I

MAS-I is a computer-based four-hour multiple-choice exam. A candidate is supposed to sit for exam MAS I only after having passed exams P and FM. The exam syllabus covers four topics with various weights. The topics and learning objectives for each topic follow.

A. Probability models (Stochastic Process and Survival Models), 20% to 35%
1. Understand and apply the properties of Poisson processes.
2. Calculate mean, variance, and probability for any Poisson process and the interarrival and waiting distributions associated with the Poisson process.
3. Calculate moments associated with the value of the process at a given time for a compound Poisson process.
4. Apply the Poisson process concepts to calculate the hazard function and related survival model concepts.
5. Calculations related to the joint distribution of more than one source of failure in a system (or life) and using Poisson Process assumptions.
6. Understand properties and calculations related to discrete Markov Chains under both homogeneous and nonhomogenous states.
7. Solve Life Contingency problems using a life table in a spreadsheet and understand the linkage between the life table and the corresponding probability models.
8. Master basic computer simulation methods.

B. Statistics, 15% to 30%
1. Perform point estimation of statistical parameters using Maximum likelihood estimation.
2. Test statistical hypotheses including Type I and Type II errors.
3. Understand insurance applications of the Exponential, Gamma, Weibull, Pareto, Lognormal, Beta, and mixtures.
4. Calculate Order Statistics of a sample for a given distribution.

C. Extended Linear Models, 35% to 50%
1. Understand the assumptions behind different forms of the Extended Linear Model and be able to select the appropriate model.
2. Evaluate models developed using the Extended Linear Model approach.
3. Understand the algorithms behind the numerical solutions for the different forms of the Extended Linear Model family, and interpret the statistical software outputs.
4. Understand and be able to select the appropriate model structure for an Extended Linear Model given the behavior of the data set to be modeled.

D. Time Series with Constant Variance, 10% to 20%
1. Use time series to model trends.
2. Model relationships of current and past values of a statistic/metric.
3. Understand forecasts produced by ARIMA.
4. Perform time series analysis using regression.

Exam MAS I sample questions and answers

Although exams MAS I and II are both in the format of multiple choice, the sample questions are presented in the format of written answers.

Question 1 *Losses follow a memoryless distribution with mean 1000. Each loss is insured and subject to a deductible of 500.*

Calculate the average insurance payment made on losses that exceed the deductible.

Answer The memoryless property of a distribution X means that $E(X - d|X > d) = E(X)$. Since each loss is subject to a deductible of 500, the insurance payment is $X - 500$ when $X > 500$ and zero otherwise.

When the average payment is calculated for losses that exceed the deductible, its value is simply $E(X - 500|X > 500) = E(X)$, therefore, the answer is 1000.

Question 2 *A building is powered by three generators with independent lifetimes, each following an exponential distribution with mean of one year. All three generators are started at the same time.*

Let T be the time in years between the first and last generator failure.
Calculate $E[T]$.

Answer Let $X_i, i = 1, 2, 3$ be the values of lifetime in years for each generator, denote their common CDF and pdf with $F_X(x) = 1 - e^{-x}, f_X(x) = e^{-x}$.

Then $T = \max(X_1, X_2, X_3) - \min(X_1, X_2, X_3)$.

Let $Y = \max(X_1, X_2, X_3)$.

$F_Y(y) = [F_X(y)]^3 = [1 - e^{-y}]^3$, and

$$f_Y(y) = \frac{d}{dy}[F_X(y)]^3 = 3[1 - e^{-y}]^2 * e^{-y} = 3e^{-x} - 6e^{-2x} + 3e^{-3x}$$

Thus $E(Y) = 3\int_0^\infty xe^{-x}dx - 6\int_0^\infty xe^{-2x}dx + 3\int_0^\infty xe^{-3x}dx$

$= 3 - \frac{6}{2^2} + \frac{3}{3^2} = 1.5 + \frac{1}{3}$.

The integration is sped up using a useful result: $\int_0^\infty t^{k-1}e^{-bt}\, dt = \frac{\Gamma(k)}{b^k}$, where $\Gamma(k)$ is Gamma function.

Let $Z = \min(X_1, X_2, X_3)$

$$F_Z(z) = 1 - P(Z > z) = 1 - \left[P(X > z)\right]^3$$

$$= 1 - \left[1 - F_X(z)\right]^3 = 1 - \left[e^{-z}\right]^3 = 1 - e^{-3z}, \text{ and}$$

$$f_Z(y) = \frac{d}{dy} F_Z(z) = 3e^{-3z}, \text{ thus } Z \text{ is exponential distribution with mean } \frac{1}{3}.$$

Therefore $E[T] = E(Y) - E(Z) = 1.5.$

Question 3 *An actuary is using the inversion method to estimate the waiting time until the fifth event of a Poisson process with a rate $\lambda = 1$.*

Five random draws from $U(0,1)$ are provided below:

| 0.2 | 0.7 | 0.8 | 0.3 | 0.5 |

Calculate the simulated waiting time until the fifth event.

Answer For a Poisson process with rate λ, the interarrival times $X_1, X_2 \ldots$ are independent and each X_i follows an exponential distribution with mean λ.

Hence the waiting time for the first five events $X_1, X_2 \ldots, X_5$ all are exponential distribution with mean $\lambda = 1$.

For each waiting time, the CDF value P_i is randomly drawn from $U(0,1)$, and the corresponding value x_i is found by $x_i = -\ln(1 - p_i)$.

So the simulated values of interarrival time are 0.223, 1.204, 1.609, 0357, and 0.693.

Therefore, the waiting time until the fifth event is $0.223 + 1.204 + 1.609 + 0.357 + 0.693 = 4.086$.

Question 4 *You are given the following information:*

- *X is a random variable from a single-parameter Pareto distribution with $\alpha = 5$ and unknown θ*
- *$\bar{x}$ is the sample mean of n independent observations from this distribution*
- *$c\bar{x}$ is an unbiased estimator of θ*

Calculate c.

Answer For single-parameter Pareto distribution, $E(X) = \dfrac{\alpha\theta}{\alpha - 1}$, thus

$\theta = \dfrac{\alpha - 1}{\alpha} E(X)$, and $\dfrac{\alpha - 1}{\alpha}\bar{x}$ is an unbiased estimator of θ.

Therefore $c = \dfrac{\alpha - 1}{\alpha} = \dfrac{4}{5}$.

Question 5 *Suppose that* $X_1, \ldots, X_{10}$ *is a random sample from a normal distribution with:*

$$\sum_{i+1}^{10} X_i = 100, \text{ and } \sum_{i+1}^{10} X_i^2 = 2000$$

The parameters of this distribution are estimated using the method of moments with raw moments only.

Calculate the estimated variance of this distribution.

Answer Using the method of moments with raw moments:

$$E(X) = \mu = \frac{\sum_{i+1}^{10} X_i}{10} = 10$$

$$E(X^2) = \frac{\sum_{i+1}^{10} X_i^2}{10} = 200$$

Thus the estimated variance is $E(X^2) - [E(X)]^2 = 200 - 10^2 = 100$.

Question 6 *You are given the following Markov chain transition probability matrix:*

$$P = \begin{bmatrix} 0.0 & 0.3 & 0.3 & 0.4 & 0.0 \\ 0.4 & 0.0 & 0.0 & 0.6 & 0.0 \\ 0.0 & 0.0 & 0.0 & 1.0 & 0.0 \\ 0.5 & 0.5 & 0.0 & 0.0 & 0.0 \\ 0.0 & 0.0 & 0.0 & 0.0 & 1.0 \end{bmatrix}$$

Determine the number of recurrent states in this Markov chain.

Answer State 5 is absorbing, thus recurrent.

States 1 through 4 are accessible from each other, thus all recurrent. Therefore all five states are recurrent.

Question 7 *You are given the following information about the distribution of losses:*

- *Losses follow an exponential distribution with mean θ.*
- *Insurance payments for each loss are subject to a deductible of 500 and a maximum payment of 30,000:*
  - *Insurance payment* = $\min[30,000, \max(0, loss - 500)$.
- *No insurance payments are made for losses less than 500.*
- *A random sample of five insurance payments are drawn:*

$$1000 \quad 4900 \quad 7000 \quad 19{,}500 \quad 30{,}000$$

Calculate the maximum likelihood estimate of θ.

Answer The density function of exponential distribution is $f(x) = \dfrac{1}{\theta}e^{-x/\theta}$, and the cumulative distribution function is $F(x) = 1 - e^{-x/\theta}$.

The five insurance payments correspond to five loss amounts:

$$x_1 = 1{,}500, \; x_2 = 5{,}400, \; x_3 = 7{,}500, \; x_4 = 20{,}000, \; x_5 \geq 30{,}500.$$

Taking into consideration that the information for losses below 500 is truncated, the likelihood function is

$$L(\theta) = \prod_{i=1}^{4} \frac{f(x_i)}{P(X > 500)} * \frac{P(X > 30{,}500)}{P(X > 500)}.$$

Now the log-likelihood function is

$$l(\theta) = \prod_{i=1}^{4} \ln\big(f(x_i)\big) + \ln P(X > 30{,}500) - 5\ln P(X > 500)$$

$$= -4\ln\theta - \frac{\sum_{i=1}^{4} x_i}{\theta} - \frac{30{,}500 + 5(-500)}{\theta}$$

Apply the first order condition:

$$l'(\theta) = -\frac{4}{\theta} + \frac{\sum_{i=1}^{4} x_i + 28{,}000}{\theta^2} = 0$$

The MLE estimate of θ is $\hat{\theta} = \dfrac{\sum_{i=1}^{4} x_i + 28{,}000}{4} = \dfrac{62{,}400}{4} = 15{,}600.$

Question 8 *An insurance company has classified claims into five categories based on their severity.*

The null hypothesis H_0 assumes the numbers of claims for Categories 1, 2, 3, 4, and 5 appear in the following ratios: 12:8:6:4:1.

In 2017, the insurance company recorded the numbers of claims as follows:

Category	# of claims
1	1172
2	829
3	605
4	347
5	102

Calculate the Chi-square goodness-of-fit statistic.

Answer The total number of claims is $1172 + 829 + 605 + 347 + 102 = 3055$.
If the null hypothesis is true, the expected number of claims in the five categories
are 1182.6, 788.4, 591.3, 394.2, and 98.5, respectively.

Thus, the Chi-square test statistic is:

$$\frac{(1182.6 - 1172)^2}{1182.6} + \frac{(788.4 - 829)^2}{788.4} + \frac{(591.3 - 605)^2}{591.3} + \frac{(394.2 - 347)^2}{394.2} + \frac{(98.5 - 102)^2}{98.5}$$
$$= 8.2791.$$

Question 9 *You are given the following information about a sample,*
$X_1, X_2, \ldots, X_n$:

- $X_1, X_2, \ldots, X_n$ *are all mutually independent*

- $X_i \sim \text{Gamma}(\alpha_i, \theta)$, *for* $i = 1, 2, \ldots, n$.

- $\alpha_i = \dfrac{1}{n}$ *for all* i

- $Y = \displaystyle\sum_{i=1}^{n} X_i$

Calculate the probability that $Y > \theta$.

Answer Using moment generating function it can be proved that $Y = \displaystyle\sum_{i=1}^{n} X_i$
follows Gamma distribution with $\left(\displaystyle\sum_{i}^{n} \alpha_i, \theta \right) = (1, \theta)$, thus an exponential with
mean θ.

Hence $\Pr(Y > \theta) = 1 - \left(1 - e^{-\frac{\theta}{\theta}} \right) = e^{-1} = 0.3679$.

Question 10 *In order to predict individual candidates' test scores, a regression was performed using one independent variable, Hours of Study, plus an intercept. Below is a partial table of data and model results:*

Candidate	Test score	Hours of study	Leverage	Standardized residuals
1	2041	538	0.6205	−1.3477
2	2502	548	0.2018	−0.4171
3	2920	528	0.6486	−1.1121
4	2284	608	0.2807	1.1472

Calculate the number of observations that are influential using Cook's Distance with a unity threshold.

Answer The formula for Cook's Distance for the *ith* observation is

$$D_i = \frac{1}{p}\left(\frac{e_i}{s}\right)^2 \left[\frac{h_{ii}}{\left(1 - h_{ii}\right)^2}\right]$$ where p is the number of independent variables

in the model (equal to 1 here), e_i/s is the standardized residual, and h_{ii} is the leverage. Using the formula, the Cook's Distance for the candidates by order are computed, and their values are 7.825, 0.055, 6.495, and 0.714. Two observations (Candidates 1 and 3) have Cook's Distance greater than 1, the unity threshold, and thus are considered influential.

Modern actuarial statistics II

After successfully passing exam MAS I, a candidate can begin to prepare for exam MAS II, which is also a computer-based four-hour exam with multiple choice questions. The exam syllabus covers four topics with various weights. The topics and learning objectives for each topic follow.

A. Introduction to Credibility, 5% to 15%
   1. Understand the basic framework of credibility and be familiar with limited-fluctuation credibility, including partial and full credibility.
   2. Understand the basic framework of Bühlmann credibility.
   3. Calculate different variance components for Bühlmann credibility.
   4. Calculate Bühlmann and Bühlmann-Straub credibility factor and estimates for frequency, severity, and aggregate loss.
   5. Understand the basic framework of Bayesian credibility.
   6. Calculate Bayes estimate/Bayesian premium.
   7. Understand Bayesian versus Bühlmann credibility for conjugate distributions.
   8. Calculate credibility estimates using the nonparametric empirical Bayes method.

B. Linear Mixed Models, 10% to 30%
1. Understand the assumptions behind linear mixed Models and evaluate how to set up a linear mixed effect model design to best accomplish the goals of the modeling exercise.
2. Understand the algorithms behind the numerical solutions for the linear mixed model to enable interpretation of output from the statistical software employed in modeling to make appropriate choices when evaluating modeling option.
3. Understand and be able to select the appropriate model structure and variable selection for a linear mixed model given the behavior of the data set.

C. Bayesian Analysis and Markov Chain Monte Carlo, 45% to 65%
1. Understand the basis and basics of Bayesian analysis and incorporate that understanding when interpreting model results.
2. Evaluate the different options available when creating and using Bayesian models for a given modeling assignment. Understand how to set up a Bayesian MCMC model and evaluate how a given set of design choices affects the results of a model.
3. Understand Bayesian computation, how Markov Chain Monte Carlo methods are used, and how to evaluate model performance. Interpret and calculate diagnostics of simulation performance to evaluate when a given modeling approach should be used.
4. Understand how to apply model checking, evaluation, comparison, and expansion techniques as an aid to interpreting and evaluating model diagnostics.

D. Statistical Learning, 10% to 20%
1. Understand the computations behind K-nearest neighbors (KNN) and be able to explain how it works in practice and its relationship with Bayes classifier.
2. Understand the computations involved in building decision trees, the purpose of tree pruning, and how extensions such as bagging, random forest, and boosting can improve the prediction accuracy of tree-based methods.
3. Understand the purpose of, and the computations behind, principle components analysis (PCA) and be able to interpret related software outputs.
4. Be familiar with the purpose of, and the computations behind, clustering procedures and be able to interpret related software outputs.

Exam MAS II sample questions and answers

Question 1 *An insurance company writes business in three states. The company experiences the following aggregate loss distributions over the past 8 years.*

State	Mean annual loss (in millions)	Standard deviation of annual loss (in millions)
U	147	72
V	57	27
W	54	24

Calculate the estimate of next year's aggregate loss for State W using the nonparametric empirical Byes method.

Answer Since the volume of business and the true underlying size of the average loss have not changed during the observation period and will continue to remain stable next year, we can apply the nonparametric empirical Bayes method with equal size.

We have

Overall mean estimated value is $\hat{\mu} = \dfrac{147 + 57 + 54}{3} = 86.$

Estimated mean of process variance $\hat{v} = \dfrac{1}{3}\left[72^2 + 27^2 + 24^2\right] = 2163$

Estimated variance of hypothetical means is

$$\hat{a} = \frac{1}{3-1}\left[\left(147-86\right)^2 + \left(57-86\right)^2 + \left(54-86\right)^2\right] - \frac{2163}{8}$$
$$= 2522.625$$

Credibility factor is $Z = \dfrac{8}{8 + 2163/2522.625} = 0.9032$

The credibility estimate of next year's aggregate loss for State W is

$$\left(0.9032\right)\left(54\right) + \left(1 - 0.9032\right)\left(86\right) = 57.098.$$

Question 2 *When modeling claim frequency, an actuary chooses to use the following Bayesian model:*

$$Y \sim \text{Poisson}\left(\mu\right)$$

$$\mu \sim \text{Gamma}(\alpha = 1, \theta = 0.1)$$

Suppose the actuary observed 10 claims from 250 policies in the past year. Calculate the mean of the posterior predictive distribution for 250 policies.

Answer Recognizing the Poisson model distribution and prior Gamma are conjugate pair, it immediately follows that the posterior distribution of μ is still Gamma distribution, and the posterior predictive distribution is negative binomial with mean equal to $\left(\alpha + \sum_{i=1}^{250} y_i \right) \left(\dfrac{\theta}{n\theta + 1} \right)$.

We are given that $n = 250, \sum_{i=1}^{250} y_i = 10, \ \alpha = 1, \theta = 0.1$, thus the mean of the predicted claim number for one policy is

$$\left(\alpha + \sum_{i=1}^{250} y_i \right) \left(\frac{\theta}{n\theta + 1} \right) = \left(1 + 10 \right) \left(\frac{0.1}{250(0.1) + 1} \right) = 0.0423.$$

Therefore, the mean of the posterior predictive distribution for 250 policies is $(250)(0.0423) = 10.58$.

Question 3 *You fit a model using MCMC and reviewed the results. You are presented with three potential problems and plausible solutions for those potential problems.*

I.   *Potential Problem: Effective sample size is small as a result of autocorrelation of samples.*
     Plausible Solution: Increase the number of iterations and thin the chain.
II. *Potential Problem: The sampling algorithm has not been adapted.*
     Plausible Solution: Reduce the number of warmup iterations per chain.
III. *Potential Problem: Your chain has converged but you are not certain that it converged in the correct region of the parameter space.*
     Plausible Solution: Increase the number of iterations.
     Determine which of the proceeding represents both a potential problem and an appropriate plausible solution to that problem.

Answer I only.

Question 4 *You are given three statements about the k-means clustering algorithm.*

I.   *The k-means clustering algorithm requires that observations be standardized to have mean zero and standard deviation one.*
II. *The k-means clustering algorithm seeks to find subgroups of homogeneous observations.*

III. *The k-means clustering algorithm looks for a low-dimensional representation of the observations that explains a significant amount of the variance.*
Determine which statements I, II, or III are true.

Answer II only.

Question 5 *You are given:*

- A dataset contains 500 observations for five predictor variables: $\{X_1, X_2, X_3, X_4, X_5\}$.
- Each predictor has been standardized to have mean 0 and standard deviation 1.
- Each of the 500 observations takes the form $\{x_{1i}, x_{2i}, x_{3i}, x_{4i}, x_{5i}\}$ for i ranging from 1 to 500.
- For each observation, a new predictor Z is calculated by projecting onto the first principal component.
- The projection for the ith observation is denoted by z_i.

The total variance present in the dataset is equal to 4 and $\sum_{i=1}^{500} z_i^2 = 750$.

Calculate the proportion of variance explained for the first principal component.

Answer Since the average of $\{z_i : i = 1, 2, \ldots, 500\}$ is zero from the design of

the PCA, the variance of first principal component is $\dfrac{1}{499} \sum_{i=1}^{500} z_i^2 = \dfrac{750}{499} = 1.503$.

Given the total variance presented in the dataset of 4, the proportion of

variance explained for the first principal component is $\dfrac{1.503}{4} = 37.58\%$.

Question 6 *An actuary is using hierarchical clustering to group the following observations: 13, 40, 60, 71.*
The actuary recalculates the clustering using two linkage methods: complete and average.

- h_{comp} is the height of the final fuse using complete linkage.
- h_{avg} is the height of the final fuse using average linkage.

Calculate $|h_{comp} - h_{avg}|$.

Answer $h_{comp} = 58$, $h_{avg} = 44$, hence $\left| h_{comp} - h_{avg} \right| = 14$.

Question 7 *A book of 122 commercial policies is observed for one year. The observed claim counts distribution is shown below:*

Claim count	Number of policies
0	45
1	32
2	27
3	10
4	7
5	1

Calculate the log-cumulative odds of three claims.

Answer

$$\log\frac{F(3)}{1-F(3)} = \log\left(\frac{45 + 32 + 27 + 10}{7 + 1}\right) = \log 14.25 = 2.66.$$

Question 8 *You are conducting an experiment with two levels: packets of pea plant seeds and the pea plant seeds themselves. You are interested in the effect of different types of light on heights of the pea plant after four weeks. Below is the final model:*

$$\text{HEIGHT}_{ij} = \beta_0 + \beta_1 * \text{TREAT1}_j + \beta_2 * \text{TREAT2}_j + \beta_3 * \text{TREAT3}_j + \beta_4 * \text{MASS}_{ij}$$

$$+ \beta_5 * \text{NUMSEEDS}_j + u_j + \epsilon_{ij}.$$

- TREAT1, TREAT2, and TREAT3 correspond to 3 of the 4 light types used.
- MASS is the mass of the given seed in grams.
- NUMSEEDS is the number of seeds in the seed packet.
- The model output is given below:

```
##Fixed effects: height ~ treatment + mass + numseeds
##
##                    Value        Std.Error      DF
##(Intercept)         10.118       0.7067         294
##TREAT1              -5.320       0.3493         22
##TREAT2              -7.824       0.3388         22
##Treat3              -4.373       0.3474         22
##mass                 0.111       0.1049         294
##numseeds            -0.053       0.0422         22
```

Determine the equation to predict pea plant height for a specific seed in a specific seed packet under TREAT1.

Answer

$$\text{HEIGHT}_{ij} = 10.118 + (-5.320) + 0.111*\text{MASS}_{ij} + (-0.053)*\text{NUMSEEDS}_j$$
$$= 4.798 + 0.111*\text{MASS}_{ij} + (-0.053)*\text{NUMSEEDS}_j + \hat{u}_j.$$

Question 9 *You are reviewing a dataset of automobile accidents using a linear mixed model.*

- Your null hypothesis is that the variance of the residuals remains constant across all four quarters (Quarter 1, Quarter 2, Quarter 3, Quarter 4).
- The alternative hypothesis is that the variance of the residuals varies by quarter.
- To test your hypothesis, you use the log-likelihood test using the Chi-square distribution.

Determine the degrees of freedom used for the test.

Answer The degrees of freedom for the null hypothesis is 1, and that for the alternative hypothesis is 4, thus the degrees of freedom for the test, which is that of the Chi-square distribution, is $4 - 1 = 3$.

Question 10 *An actuary is constructing a credibility formula to apply to claim severity. To visualize the partial credibility, a plot showing credibility as a function of claim counts is created.*

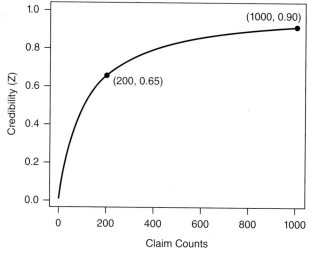

Determine which credibility method the actuary is using.

A. Limited-fluctuation credibility with a full-credibility standard of 1300 claims.
B. Limited-fluctuation credibility with a full-credibility standard of 800 claims.
C. Limited-fluctuation credibility with $\lambda_F = 541$ and $C_X = 1.053$.
D. Bühlmann credibility with $EVPV = 0.266$ and $VHM = 0.016$.
E. Bühlmann credibility with $EVPV = 0.844$ and $VHM = 0.008$.

Answer Limited-fluctuation credibility

$$Z = \sqrt{\frac{\text{\# claims available}}{\text{\# claims needed for full credibility}}}$$

Model A. $Z = \sqrt{\dfrac{200}{1300}} = 0.39$ for $n = 200$ while the given credibility is
0.65. Wrong model.
Model B. The number of claims needed for full credibility is 800. When $n = 1000 > 800$, Z should be 1, not 0.9. Wrong model.

Model C. The values $\lambda_F = 541$ and $C_X = 1.053$ are for aggregate claims, not applicable to claim severity. Wrong model.

Bühlmann credibility $Z = \dfrac{n}{n + EVPV/VHM}$.

Model D. For $n = 200$, $Z = \dfrac{200}{200 + 0.266/0.016} = 0.92$, not equal to the
given 0.65. Wrong model.

Model E. $Z = \dfrac{n}{n + EVPV/VHM} = \dfrac{n}{n + 105.5}$. We can verify that the two
credibility values computed based on the formula are the same as those given. Model E is the correct one.

Chapter 3

Actuarial jobs

Actuaries primarily provide actuarial services. Given their role in the society worldwide, it is not surprising that a web search for "actuarial services" may produce over 41,200,000 hits (it was 200,000 hits at the time this book's first edition was published 20 years ago). Any such search shows that actuarial employers come in all sizes, large and small. Actuaries work not only in consulting firms and insurance companies but also in government departments, in the human resource offices of most major companies, and in small firms specializing in a variety of actuarial tasks.

It is difficult to present an accurate snapshot of the state of the actuarial world at any given moment in time. Globalization of economies and changes in financial regulations and structures often blur the distinction between the actuarial and nonactuarial roles of many companies. In this text, therefore, we concentrate on representative companies, large and diversified enough to illustrate the spectrum of actuarial careers. You will find basic information about some of the top employers of actuaries. The list does not include alternative forms of employment in business, banking, teaching, human resource management, administration, government, and so on. For more detailed information on actuarial careers elsewhere, we refer you to the websites listed in Appendix D at the end of this book. The profiles vary from company to company to illustrate different aspects of actuarial employment.

As is the case in many professions, social networking has become a significant component of job search and professional advancement. We illustrate the role of social networking in the lives of actuaries by including some selected but somewhat random search results. This will give you an idea of how you might go about developing your own professional social network, both as an actuarial student and as an established actuary.

We begin by answering some of the basic questions you may have about getting started with your actuarial career.

Actuaries' Survival Guide, Third Edition. DOI: 10.1016/B978-0-443-15497-3.00003-6

3.1 Landing Your First Job

Q: What advice and tips would you give, based on your participation in recruiting activities from an employer's point of view, to students applying for an actuarial position in your company?

ANS: Your technical skills are definitely noticed (ability to use Microsoft Excel, SQL, for example). But impressions carry more weight to certain team leaders. We have an undefrstanding that actuarial students will be able to develop technical skills and we don't expect them to be experts early in their career. Most of the time, we are looking for someone who will fit in with the culture of the team. Someone who we will get along with, and has the soft skills (time management, team player mentality, etc.) necessary to be successful in a professional setting.

ANS: Gauging their willingness to learn and how agile they can be when faced with different situations.

ANS: I think the student would be a strong candidate if he/she passes 2 to 3 exams, is familiar with one or two programming languages such as SAS/R/Python/SQL, has an internship on relative work, has statistical experiences either in work or in school courses, is able to answer questions related to the resume and is able to have a good conversation during the interview.

ANS: Pass exams, network, and get referrals.

ANS: I think understanding real-life actuarial concepts is important. In an interview, I do not want to hear about the exams passed. I want to hear about how you can add value to the team through your knowledge or ability to learn.

ANS: I will advise students applying for an actuarial position at BHSI that they should try and pass at least three exams and should have a full understanding of Microsoft Excel and its applications. Knowing many Microsoft Excel shortcuts helps an actuary to get work done faster and more effectively. Having a strong understanding of SQL and R-studio and being able to communicate effectively orally, written, and analytically.

ANS: I assume this is for entry-level jobs. First, start with an internship. If you don't have any actuarial internship, it's really hard to get an actuarial position. During the interview, show the interviewer what your technical skills are and how good you are. Give details of what you did in projects, but not the confidential information related to your company. Be prepared for the regular questions, like why you want to become an actuary or why you chose our company. Don't exaggerate or lie about your skills. If you just know the basics of R, don't say that you build a model in R. Interviewers can tell if you are really good or not.

ANS: Companies nowadays put more weight on how candidates demonstrate team collaboration.

ANS: 1. Pass Exams! 2. Learn Microsoft Excel and Python. 3. Work in teams and demonstrate the ability to communicate and learn.

ANS Have at least 2 to 3 actuarial exams passed; have internship experience in the industry if applying for a full-time job; decent GPA; be confident and enthusiastic.

ANS: Insurance knowledge, internships, 2 to 3 passed exams, and good programming skills (VBA or R).

ANS: For entry-level positions, it is important for students to demonstrate their ability to ask good questions during the interview process. Some questions are an opportunity for the student to demonstrate their knowledge; others, especially good follow-up questions, help to demonstrate their ability to learn quickly in a fast-paced environment.

ANS: Demonstrate that you are able to hold a conversation in both a nontechnical and technical forum.

ANS: Emphasize communication skills and have an internship and 1 to 2 exams under your belt. Be a good listener and speaker and be someone people want to work with. The technical and analytical skills will come with the job-you have to get hired first, and that comes with being personable.

ANS: Make sure your resume is only one page long. Also, be able to explain in detail anything that is put on your resume.

ANS: From an employer's point of view, the key aspects I look for are hardworking, a great team player, and willing to be open to new projects.

3.2 Company Reputation

Q: Are there companies with a good or bad reputation in the actuarial industry? Give some examples and explain why.

ANS: I think this depends on what your goals are as an individual. Some people have bad opinions of consulting as that doesn't suit their desires for a career. Others have poor opinions on carriers[1] due to the work being "boring." It all depends on the individual.

ANS: I am not aware of any.

ANS: I only have been working in AEGIS and the Auto Club Group, both are very good companies.

ANS: Consulting companies have bad work-life balance due to long hours during busy season. Insurance companies are known to have a better work-life balance.

ANS: I think consulting has a tough reputation with the workload. It is difficult to pass exams when working long and grinding hours.

ANS: I do not have enough knowledge about companies with a good or bad reputation in the actuarial industry, so I cannot speak to this question.

ANS: Most companies are similar. If you are on a good team, you will love this company and stay there for a long time; but if your team doesn't help you grow or treats you badly, you might want to leave ASAP. I know a company where people are ruthless and most don't care about you, and entry-level actuaries do dirty work with no promotions. Another company is known to be disorganized. I heard that some people left after less than six months of work. I can name one good company: Guy Carpenter. My manager used to work there and she liked it.

ANS: I cannot think of any companies with a bad reputation. Larger companies have a good reputation for having a rotational program, which I highly recommend for new students.

ANS: There certainly are, but I'm not so familiar with them. My suggestion is to ask about the company/department culture during an interview and find out if it is the right fit for you.

ANS: Consulting companies work you more while insurance companies have a better work-life balance.

ANS: I don't believe so.

ANS: Some companies are known for their strict student programs-forced job rotations every few months, mandatory study time and exam results, etc. Other companies are more relaxed with their study programs. Overall, actuaries tend to act with integrity so there is not much concern for companies with bad reputations.

ANS: Yes, I don't know of any companies specifically, but there are some companies that overwork actuaries and don't give ample study hours or ample pay increases for passing exams.

[1]Carriers refer to insurance companies who carry risks transferred from people and organizations that purchase insurance coverage.

3.3   Salary and Benefits

Q: **What are the responsibilities of new employees in actuarial entry positions in your company, and what are their typical tasks and salary ranges?**

ANS: Typical starting salary may be around $65 to $75 K[1] now (though I'm not sure). Tasks are really getting to know the business, working on basic pricing, or reserving projects. Nothing crazy, no decision-making on development factors or trend rates (that's for more senior members to decide). We are mostly looking for new members to update the exhibits, assess new data for reasonability, and give us their initial thoughts/observations.

ANS: Analyze data, develop programs, and prepare client presentations. Not sure of latest salary ranges.

ANS: New employees are responsible for some simple tasks and easy work, and also participate in a big project with a mentor or manager's lead. Typical tasks include rate revision and data analysis assignment. Salary ranges from $65 K to $70 K a year.

ANS: They help populate worksheets and models to generate results. Salary ranges from $65 K to $70 K for new graduates.

ANS: I think our new hires make around 80,000 (dependent on tests). They are expected to start with no knowledge, but are expected to gain some knowledge quickly.

ANS: New employees at the reserving team at BHSI provide supporting roles in all areas such as quarterly reserve closings, run data checks and data validations, clean dataset, by identifying incomplete dataset, learn about the business, support in year-end actuarial report by converting documents to pdf format, help in proofreading of documents, and provide enough hands-on opportunity to learn about business partners and much more.

ANS: Entry-level actuaries are doing things like price monitoring, data cleaning and mapping, profitability analysis, updating templates, etc. Salary ranges are $60 K to $70 K, based on how many exams you passed.

ANS: Ad hoc reporting, supporting senior actuaries on various projects.

ANS: Entry-level associates are responsible for completing smaller tasks such as updating parts of an actuarial model, repeating a process that was performed last year. I recommend DW Simpson or a similar agency for salary data.

ANS: Salary range is about $65 K to $80 K depending on the number of exams passed. Typical tasks in the pricing team include routine processes such as preparing experience reports and meeting slides, running basic pricing models, market research, competitive analysis, etc.

ANS: Updating workbooks, presentations, and initial pass of reserve reviews. Starting salary is about $65,000.

ANS: New employees are usually expected to take over routine processes while learning various concepts for the work.

ANS: Data cleansing, mapping exposure information, documenting process, and providing analyst support to project. I would say a new actuarial student with two exams would be around $75 K to $80 K.

ANS: New entry-level actuaries at my company tend to spend their first few months learning about the company and figuring out where they can bring most value to the team. We don't have a formal rotation program, so actuaries are simply hired into a specific role – they learn, on the job, how to do what's required of them. Typically entry-level actuaries start at $70 K to $75 K a year at my company, depending on the number of exams they've passed. And this number may have increased in recent years, due to inflation.

ANS: Responsibilities for new employees are very basic and rudimentary. It is important to first give entry-level actuaries exposure to how pulling data works and the different products offered by the company. Typical tasks are monthly reports on trends, membership growth/decline, etc. Salary ranges vary by area.

ANS: New employees are responsible for reserving one business unit each quarter, as well as updating various exhibits in the off-quarter months. Salary range for an assistant is $60 K to $65 K.

[1]All monetory values in the survey responses are in US dollars.

Q: How do you negotiate your salary, and have you always been satisfied with the salary and benefit packages you have had? Illustrate your answer from your knowledge of the practice of specific companies.

ANS: I negotiated up a very small amount out of school. Mostly, if you're unsatisfied with your compensation, it may be time to re-evaluate your career and look for a change with a new company. The only time I've been less satisfied with my compensation was when I took another opportunity that better suited my career goals anyways. There are annual studies on actuarial compensation so you can always compare your current situation versus others.

ANS: My salary is benchmarked with the industry, and we do have a cohort system these days where your title and level designate the salary you will receive, hence eliminating the possibility of discrimination based on gender or race.

ANS: Based on my work content and my current exams passed, I'm satisfied with the salary and benefit packages I have. I asked for a promotion after one year of working, and I was promoted after an assessment interview. I think it would be better to negotiate with your competence and specific contribution you made to the company.

ANS: I did not negotiate my salary as I did not have much leverage. Switching companies always produces the highest raise. I was not satisfied with my total compensation as an actuary.

ANS: I think knowing your value is important. But I also think you want to work at a company that values you. I have never been in a position where I felt underpaid/valued because my manager is most vocal on my behalf.

ANS: I have not been in a position yet to negotiate my salary. I know that my colleagues in the industry do get salary adjustment and bonuses due to successful meeting of benchmarks.

ANS: I am satisfied with my salary and benefits. It's quite standard. We have salary raises every year. The more exams you are able to pass, the more you get paid.

ANS: I'm satisfied with my contributions being recognized.

ANS: Understanding the market landscape is an essential tool for negotiation. Also, some industries allocate more money for talent than others.

ANS: Not really. My entry-level job package was pretty much industry standards, and after that, the salary increase with each passed exam is specified in the program. Annual bonus/performance salary increase are not the highest in the industry, but I'm satisfied overall.

ANS: Salary raises have been satisfactory, so I have not needed to negotiate.

ANS: For the first job, salary negotiations are likely not the priority, but to evaluate the opportunities to develop and learn. I think it helps in salary negotiation to lay out the request with quantified support. Although there is always a risk for an offer to fall through, my experience is that after a candidate is selected, it is likely fine for a candidate to negotiate the package within reason.

ANS: I had always taken the route to get more on the salary instead of the sign-on bonus/retention bonus, since that number isn't a one-time figure and will follow you through your career.

ANS: I have always received an annual raise when performance reviews came around. Sometimes it was a little lower than I could've probably negotiated for, but I don't like to penny-pinch and risk my job over a couple thousand dollars – if I was going to do that, I would want to have an offer from another company and use that as leverage. Plus, my company has very good benefits that I understand also makeup part of my salary – even if I got a higher base pay elsewhere, I may not get the same 401(k) match or health insurance. I negotiated my salary for the first (and only) time when I changed roles, and I think my success in that negotiation was due to the fact that I was moving from a position where I worked a lot of overtime (and got OT pay for it) to a position where I would be salaried (so no more OT pay). I calculated my equivalent salary as if I were to become exempt from OT in my previous role, and used that number to ask for an additional $5 K in the new role.

ANS: I negotiate my salary at the semi-annual performance review. I have always been satisfied with my salary and benefits.

ANS: I'm satisfied with my salary and package. I would recommend doing your research before applying.

ANS: I'm satisfied with my contributions being recognized.

Many employers have actuarial training programs for actuarial students. A typical training program will include assistance with exam preparation, rotational assignments to expose the student to different aspects of actuarial work, seminars, purchase of study material, paid study time, and reimbursement of the costs associated with the writing of the examinations. The survey illustrates the value of such training programs for career development.

◆◆◆

Q: Based on your employment experience, describe the support different companies give to students to prepare for actuarial examinations (study days, payment of examination fees, purchase of study materials, etc.)

ANS: Generally, companies are very generous with study time. The team leaders have all been there as a student, so they understand what we are going through. Most companies pay for exam fees and study materials. They give you the appropriate time off to take the exams and study. It is easy to get caught up in work and not take study time. At the end of the day, it's on you to manage your workload, manage your time, and communicate with your manager.

ANS: My company reimburses for exam fees and materials, provides study days, and gives you raises and bonuses on the exams you pass.

ANS: At my current company, for each exam, 100 to 110 hours are given for the first attempt, exam registration fee is paid, and one study forum/ textbook is covered.

ANS: Study days, payment of examination fees, purchase of study material, raise for passing.

ANS: Companies pay for all study material and exam fees. Different amount of study hours for different exams.

ANS: My company provides 3 study days per 1 hour of exam time. The first two sittings are fully reimbursed. The study material is reimbursed for the first sitting.

ANS: BHSI provides all employees preparing for actuarial examinations study hours, pays all examination fees for three sittings per exam, reimburses all the costs of study materials, and provides salary adjustment after submission of SOA/CAS exam grade.

ANS: My company gives actuaries three sittings for each exam. After three sittings, you are out of the program but you can still use your own time to study. The company only pays for a one-time purchase of study material.

ANS: Three days per exam hour plus the day before the exam to prepare for the exam. Full reimbursement if the actuarial student passes the exam.

ANS: I worked for two companies. The life insurance company has a rotational program, and the support is better compared to the pension consulting company. That's thanks to the rotational program.

ANS: Most larger companies have well-established actuarial student programs, in which they specify the number of study days for each exam, fee reimbursements, purchase of study material, the least number of exams to be passed each year to stay in the program, salary increase, etc. My company gives 3 study days for each exam hour, exam fee/ study material reimbursements, and salary increase.

ANS: Study hours – 120 hours for a 4-hour exam; exam fees and study materials are covered. If an exam is not passed by the third attempt, then these benefits are reduced.

ANS: Generally, I think insurance companies care more about the exam progressions and provide more support than consulting firms.

ANS: Study days are usually 3 days per hour of the exam, study materials paid for, and exam fees fully paid for.

ANS: I have only ever worked at one company, but in multiple roles. Every manager I've had has told me that the exams are good, but they are in no rush for me to get my ASA or FSA. They left it up to me to decide my pace and progress. They were happy to support me by making sure I get my study hours when I need them, and they reimbursed me for my exam materials and fees.

ANS: Companies usually pay for the study material and exams for the first attempt, as well as full allotment of study hours. From my experience, usually the rule is 3 study hours per hour of the exam.

ANS: Both in insurance and consulting companies, students get study hours, books, and pay increase when passing exams.

3.4   Moving Up the Ladder

Q: How fast can an actuary expect to climb the ladder in your company? Illustrate your answer and compare it with examples from other companies you may be familiar with.

ANS: This will depend on the unique situations of your team. Larger companies have more scheduled promotions, I would say, compared to smaller ones. Some companies have very specific, objective evaluation metrics while others are more subjective. This will also depend on exam progress.

ANS: Their success in the exams usually dictates how fast they climb the ladder.

ANS: One promotion a year or two years. It depends on the exams passed and workload taken.

ANS: Senior associate in 3 years, Manager in 5 years.

ANS: Usually the timeline is 2 to 3 years depending on performance and exams. You cannot move to be an "associate" until you receive ASA. Similarly, you cannot be an "actuary" until you receive FSA.

ANS: At our company, actuaries are given opportunity to grow by being available and open-minded for new areas of growth since our company is a growing company and every day new opportunities and positions are being created. Any new positions are advertised internally first and talented employees are recruited first before they go public.

ANS: My company isn't big enough so there aren't too many positions. For the entry-level actuaries, if you are okay with the salary and happy with the team, you can stay a little bit longer to (1) pass all the exams, and (2) learn not only yours, but other's lines of business.

ANS: At the early stage of an actuary's career path, I think the most important thing is to take actuarial exams and set a solid foundation. When the actuary starts to move up to the management positions, soft skills, as well as leadership skills, start to play a more important role. In general, it takes around 2 to 3 years to become an associate, and then a few more years to manager and director levels. Personally, I think the fastest way to climb up the career ladder is to stay out of the comfort zone, always stay challenged, and switch companies when better opportunities come up.

ANS: It took two years to get the title of Actuarial Analyst II. It depends on how good you are at work and also how many exams you are passing.

ANS: I am new to my current company, but my general sense is that it is hard to climb significantly faster in the first 7 to 10 years. After an actuary is seasoned, it will likely require a willingness to change jobs or employer to climb faster than expected.

ANS: I think the process is a fairly slow one, just because you attain your fellowship quickly does not necessarily guarantee you the trust of leading projects. I would say the same is with the three Brokerage (Marsh/Aon/Willis).

ANS: At my company, entry-level actuaries can expect their first promotion shortly after becoming an Associate. After that, it entirely depends on their job performance and if they become an FSA. I received my first promotion when I changed roles after getting my ASA, which took about 4 years after starting my job. I have a coworker who received their ASA in December, and they received their promotion slightly before the official ASA announcement (as they knew it was coming). It took them about 6 years. Another coworker received his ASA at the same time as me, and he probably will not be promoted for another year or so since he only has 2 years of experience. I think this is a little slow compared to other companies – I remember seeing many of my peers being promoted before they became Associates.

ANS: An actuary can expect to be promoted every 2 to 4 years, depending on the quality of his/her work. For example, I came into my company as an analyst, and was promoted to senior analyst 3 years later.

ANS: The key is to communicate with your coach and ask them to help you get to where you want to be.

3.5   Consulting vs. Insurance

Actuarial careers can be looked at in several ways. Consulting versus insurance is one of them. Here are some comments from respondents to the survey about the advantages of these career options:

Q: **What are the main advantages of working for (a) a consulting firm and (b) an insurance company?**

ANS: Advantages of working for a consulting firm include more collaboration, working with more people, seeing a wider range of projects, and communicating with clients.

ANS: (a) Learning curve is quicker, exposure to different types of projects, job likely more secure. (b) More time to study for exams, better hours in general.

ANS: Consulting firms have relatively higher salaries but people are typically busier. An insurance company's work is slower, and salary is lower compared to consulting firm.

ANS: Consulting – more money, more exposure to different areas of the insurance industry. Insurance – more predictable hours, rotation programs, easier to pass exams.

ANS: An insurance company allows you to have adequate study time. A consulting firm allows you to gain work experience.

ANS: As actuary, working for an insurance company is better than working for a consulting firm because in the insurance company, actuaries have array of opportunities to use their experiences and skill set whereas working for a consulting firm, skillset and knowledge may be limited to specialty or specific areas of insurance.

ANS: Consulting firms are flexible on time, you can work anywhere or anytime as long as you finish projects on time, but consulting firms can be very stressful; for example, a client needs an actuarial report for shareholders' meeting tomorrow, you need to get it done ASAP and send it to your manager for review. Insurance companies are less stressful on that point. The work-life balance is good.

ANS: I'm working in a life insurance company. I'm exposed to mainly product-wise problems.

ANS: Insurance companies tend to provide more work-life balance, which allows young actuaries to spend more time studying for exams.

ANS: In a consulting firm, actuaries get exposure to various kinds of actuarial functions and different product types from different types of companies. Even working on a single subject, consulting actuaries get exposed to almost all kinds of situations experienced by the industry in a very short period of time via working with different companies. Actuaries who enjoy working with clients and solving problems might find working in a consulting firm a good fit. On the other hand, when working in an insurance company, actuaries are more focused on only one or several fields of actuarial functions but are more familiar with the company-specific circumstances and are great assets to the company.

ANS: The main advantage of working for a consulting firm is the vast array of projects and problems that the actuary is exposed to. Additionally, there is an emphasis on softer skills like communication, time management, and prioritization. The main advantage of working for an insurance company is a structured career path and a focus on honing a specialty.

ANS: The main advantages of consulting are the range of projects to work on, as well as higher initial salary. Disadvantages of consulting are difficulty in finding company time to study for exams, as well as longer hours on average than working for an insurance company. Insurance advantages are a better work-life balance and more time to study for exams. Insurance disadvantages are lower initial pay and less variety of projects.

ANS: (a) More pay; (b) better work/life balance; able to spend more time on exams.

ANS: Consulting firms tend to provide more of a breadth of understanding versus depth. Consulting jobs also would generally provide more opportunity for networking. Working in an insurance company tends to provide more work-life balance and exam support.

ANS: Consulting firm work is more dynamic, and I think you get to touch on a lot more stuff as compared to working at an insurance company. I think the insurance company does provide an easier work environment to pass actuarial exams.

ANS: I have only ever worked for an insurance company. My understanding is that consultants make more money but work much harder – longer hours and more stressful assignments. Consultants also give less emphasis to exams, so many actuaries will start their careers at insurance companies, pass all their exams, then get a job as a consultant and take a substantial raise. The work-life balance also tends to be better at carriers.

ANS: The advantages of working at a consulting company are exposure to different projects and you have more opportunity to grow. In an insurance company, you work shorter hours than a consulting company and you really work on consistent projects.

Appendix A

Consulting firms

Society of Actuaries (SOA) and Casualty Actuarial Society (CAS) are two professional organizations that represent actuaries in the United States. Both organizations publish data about their memberships. The majority of about 9100 CAS members work for three types of employers: property and casualty insurance companies (53.8%), consulting companies (15.3%), and companies serving insurance businesses like rating agencies (12.5%). According to a survey conducted in 2020 of 23,000 SOA members who reported their industry, 38.03% worked in life insurance industry, 31.9% in consulting, and 13.3% in health insurance.

AM Best, the world's largest credit rating agency specializing in the insurance industry, published rankings of top actuarial firms in its publication *Best's Review* magazine in December 2022. Actuarial firms are third-party actuaries who provide an annual statutory actuarial opinion regarding an insurer's policy and claim reserves, and rankings are based on the loss reserves held by their client insurance companies that file with the National Association of Insurance Commissioners.

Medical insurance plans and retirement plans are employee benefits commonly offered by employers across the world, thus human resource consulting firms hire actuaries to provide advisory services on such plans. Zippia, a company specializing in career services, published a list of the 10 largest HR consulting firms in the world by revenues in 2020. Another career-building platform, Firsthand, asks consultants to name the best three companies in their specialization area but not their employer. The following company profiles are chosen based on the three sources described above and the author's personal knowledge.

Job descriptions included with some company profiles do not indicate that these jobs are still available. They are meant to illustrate different aspects of actuarial life, and to give real-world examples of the main theme of this guide, which stresses the bond between mathematics, business, and statistics upon which actuarial careers are built. Useful information about all of the companies below can also be found on the Facebook and LinkedIn websites of these companies.

Actuaries' Survival Guide, Third Edition. DOI: 10.1016/B978-0-443-15497-3.00009-7

A.1   Aon plc

- Company website: www.aon.com
- Careers: jobs.aon.com
- Facebook: www.facebook.com/Aonplc
- LinkedIn: www.linkedin.com/company/aon
- Twitter: twitter.com/aon_plc

Aon plc (NYSE: AON), a Fortune 500 company, is a leading global professional services firm providing a broad range of risk, retirement, and health solutions. Over 50,000 employees in 120 countries in the Middle East, Africa, Asia-Pacific, Europe, and North America empower results for clients by using proprietary data and analytics to deliver insights that reduce volatility and improve performance. The company, through its subsidiaries, offers a range of commercial risk solutions, reinsurance solutions, retirement solutions, health solutions, and data and analytic services. Aon is headquartered in Dublin, Ireland.

Actuarial careers at Aon

Opportunities at Aon for actuaries: Aon provides reliable and consistent actuarial services for domestic-only and global- defined benefit (DB) plans. Aon actuaries help multinational companies manage their plans, with a focus toward the key areas of design, financing, and operations.

Aon's Actuarial & Analytics practice is one of North America's largest Property & Casualty (P&C) consulting firms, providing actuarial services for more than 20 years. The A&A team leverages advanced analytics, modeling, and quantification expertise to services clients. Our team also publishes six industry-leading loss cost benchmarking reports which help inform our clients, and the industry as a whole, on the state of the market. These reports can also be individually tailored to help clients see how their experience compares to the competition and help them identify areas where improvements can be made to help optimize their total cost of risk.

Aon's actuarial and analytic specialists analyze past experience and evaluate potential losses critical to effectively manage risks. Our consultants evaluate the financial impact of current economic, legal, and social trends on future events.

Aon's actuarial team provides services including, but not limited to:
- Loss Reserving and Forecasting
- Collateral Review and Negotiation
- Risk Appetite Modeling and Retention Analysis
- Industry Loss Rate Benchmarking
- Cyber Modeling
- Captive Feasibility Studies

A.2 Deloitte

- Company website: www.deloitte.com
- Careers: https://apply.deloitte.com/
- Facebook: www.facebook.com/DeloitteUS (for US operation)
- LinkedIn: www.linkedin.com/company/deloitte
- Twitter: twitter.com/DeloitteUS (for US operation)

Deloitte provides industry-leading audit and assurance, tax and legal, consulting, financial advisory, and risk advisory services to nearly 90% of the Fortune Global 500® and thousands of private companies. The company's more than 345,000 professionals deliver measurable and lasting results that help reinforce public trust in capital markets, enable clients in more than 150 countries and territories to transform and thrive, and lead the way toward a stronger economy, a more equitable society, and a sustainable world.

Actuarial careers at Deloitte

Deloitte's insurance group brings together specialists from actuarial, risk, operations, technology, tax, and audit. These skill sets, combined with deep industry knowledge, allow the company to provide a breadth of services to life, property and casualty, reinsurers, and insurance broker clients.

Deloitte's actuarial consultants provide a full range of reserving and advisory services to clients. They help clients combine tax data, processes, technology, and people in ways to uncover valuable business insights and arrive at smarter solutions to business challenges.

Organizations are challenged with new emerging risks as they seek profitable growth and attempt to maximize the return on capital by understanding the risk-adjusted opportunities to deploy the capital. Deloitte's global Actuarial, Rewards & Analytics (ARA) practitioners help business leaders make informed decisions to grow revenue, manage risk and capital, reduce operational costs, and design compensation and reward programs to address critical business, financial, and insurance challenges.

Deloitte's Life Insurance Actuarial practice helps insurers stay ahead by providing industry-leading insights and solutions to address a variety of challenges. The team services across a broad spectrum of functions—actuarial, risk, sales, and corporate—and takes advantage of Deloitte's Risk, Advisory, Technology, Human Resources, and Finance practices to offer out-of-the-box ideas along with the depth and breadth of resources to clients.

A.3 Ernst & Young Global Limited

- Company website: www.ey.com
- Careers: careers.ey.com
- Facebook: www.facebook.com/EY

- LinkedIn: https://www.linkedin.com/company/ernst-young-consulting-services/
- Twitter: twitter.com/EY_US (for US operation)

Ernst & Young Global Limited, trade name EY, is a multinational professional services partnership headquartered in London, England. Member firms located in over 150 countries provide services to clients through more than 280,000 employees. Ernst & Young provides audit, tax, business risk, technology and security risk services, and human capital services worldwide.

Actuarial careers at EY

The insurance industry faces a number of challenges: technology-driven disruptions, evolving consumer expectations, intense cost and competitive pressures, profound regulatory change, unprecedented opportunity in emerging markets, environmental sustainability, and lingering economic uncertainty. EY's global team of insurance professionals combines industry experience and technical knowledge to help insurers address these pressing issues. From innovative consulting services to tax and audit advice, EY consultants help insurers to integrate technology, optimize customer experience, develop new products, create M&A strategies, adopt new business models, and address shifting workforces in order to transform and drive long-term growth.

EY's offices across the world set up various actuarial practices such as Actuarial robotic, Actuarial transformation and offshoring, Analytics and Audit of actuarial items, banking and capital market quantitative consulting service, cyber risk management and insurance, longevity, pension and pension fund advice, finance transformation in insurance, underwriting transformation and claim transformation.

A search at EY's career page with the keyword "actuarial" in June 2022 returned 180 positions in Argentina, Australia, Belgium, Bermuda, Canada, Cayman Islands, China, Cyprus, Greece, India, Ireland, Italy, Luxembourg, Malaysia, Malta, Mexico, Netherlands, New Zealand, Norway, Pakistan, Philippines, Poland, Portuguese, Singapore, South Africa, South Korea, Switzerland, Thailand, UK, and the United States. Job functions of the openings include:

Insurance and Actuarial Advisory Services—Life and Health
Insurance and Actuarial Advisory Services—Property and Casualty
Insurance Risk and Regulatory Advisory
Business Consulting—Actuary Manager
Data Science—Business Modeling and Analytics
Nonlife Actuary Manager
Director—Actuarial services
Manager—Pension and Benefits Actuarial Services
Pension Actuary
Quantitative Investment—Actuarial Services

Risk and Actuarial Digital Strategy—Consulting
Financial Service Risk Consultant

A.4 KPMG

- Company website: kpmg.com/us/en.html
- Careers: home.kpmg.com/careers
- Facebook: www.facebook.com/KPMG
- LinkedIn: www.linkedin.com/company/kpmg
- Twitter: twitter.com/KPMG

KPMG is a global network of professional firms providing audit, tax, and advisory services. KPMG firms operate in 144 countries and territories, and collectively employed more than 236,000 people in 2021, serving the needs of businesses, governments, public-sector agencies, not-for-profits, and through the audit and assurance practices, the capital markets.

Actuarial careers at KPMG

The rise of black swan events and catastrophic losses are placing the insurance industry under greater pressure than ever before. To address these shifting realities and maintain profitability, insurers must price their products more effectively, build more sensitive risk models, improve dynamic capital adequacy testing (DCAT), and enhance operational efficiency. Financial services organizations as a whole must also take steps to stress test their balance sheets and quantify their risk exposures. Achieving these goals calls for more complex and precise actuarial analyses. Widely used in the insurance and finance industries, actuarial analysis is entering into many other areas as its power and flexibility become more widely appreciated. Increasingly, actuarial skills are being sought to support multidisciplinary approaches to important business decisions.

KPMG's Actuarial practice helps firms balance business goals with financial risk by combining technical knowledge and rigorous processes with wide-ranging commercial and market experience to provide progressive, high quality, and flexible advice. The team works across a wide range of areas, including risk management, regulatory, tax, advisory, and accounting lines, providing the full breadth of services to clients.

Actuaries can look for jobs within these teams:
Pensions and other postemployment benefits insight
Risk management
Investment management
Asset management
Insurance deal advisory
Insurance accounting change

A.5 Marsh & McLennan Companies

- Company website: marshmclennan.com
- Careers: careers.marshmclennan.com/global/en
- Facebook: www.facebook.com/MarshMcLennanCareers/
- LinkedIn: www.linkedin.com/company/marshmcLennan
- Twitter: twitter.com/MarshMcLennan

Marsh & McLennan (NYSE: MMC) is a global professional services firm, providing advice and solutions to clients in the areas of risk, strategy, and people worldwide. It operates through four major businesses—Marsh and Guy Carpenter specialize in the segment of Risk and Insurance Services, while Mercer and Oliver Wyman service in the Consulting segment. The Risk and Insurance Services segment offers risk management services, such as risk advice, risk transfer, and risk control and mitigation solutions, as well as insurance and reinsurance broking, catastrophe and financial modeling, related advisory services, and insurance program management services. This segment serves businesses, public entities, insurance companies, associations, professional services organizations, and private clients. The Consulting segment provides health, wealth, and career consulting services and products, specialized management, as well as economic and brand consulting services.

Altogether, the firm employs more than 83,000 employees and serves clients in 130 countries. It is one of the Fortune 250 companies with reported annual revenue of $19 billion for 2021.

Actuarial careers at MMC

A search at the company's career webpage with the keyword "actuarial" in June 2022 returned 185 openings in various job functions: actuarial and underwriting (118), consulting (39), customer services (18), client management and sales (4), operations (3), marketing and communication (2), facilities and business services (1). The job titles included actuarial analyst, actuarial consultant, consulting actuarial analyst, health and benefits consultant, pension actuarial consultant, retirement actuarial consultant, retirement actuarial analyst intern, governmental actuarial consultant, consultant in life insurance, consultant in property and casualty insurance, reinsurance leader, senior plan analyst of employee health and benefits.

These positions are located in Belgium, Canada, France, Hong Kong, India, Ireland, Japan, Mexico, Netherlands, Poland, Portugal, Singapore, Switzerland, Taiwan, Turkey, the UK, and the United States.

A.6 Milliman, Inc

- Company website: www.milliman.com
- Careers: careers.milliman.com

- LinkedIn: www.linkedin.com/company/milliman
- Twitter: twitter.com/millimaninsight
    twitter.com/millimanhealth

Milliman is an independent risk management, benefits, and technology firm with offices in major cities around the world, headquartered in Seattle, Washington. Milliman serves the full spectrum of business, finance, government, union, education, and nonprofit organizations. Milliman's body of professionals includes actuaries, technologists, clinicians, economists, climate and data scientists, benefits and compensation experts, and many others.

Actuarial careers at Milliman

Milliman's major business branches include health, insurance, retirement and benefits, and risk, all of which offer career opportunities for actuaries. The Actuarial Consulting Solutions provides services in pricing and plan design, reserving and financial reporting, claim experience and financial analysis, and regulatory filings. Milliman's insurance products include financial modeling and industrialization, premium comparison platform, reserving and liability modeling, Solvency II reporting and compliance, insurance risk assessment, and retirement income security evaluation, among others.

As of July 1, 2022, on the company's career webpage, there were 186 actuarial listings in the title actuarial intern, actuarial analyst, actuarial analyst—property and casualty, actuarial associate, actuarial data analyst, actuary (ASA/FSA), actuarial student, actuarial and risk management consultant, commercial healthcare pricing actuary, actuarial software analyst, nonlife actuarial intern, consulting actuary, quantitative analyst, and compliance consultant—life and annuity. These positions are available in Australia, Belgium, China, India, Indonesia, Luxembourg, Malaysia, Netherlands, Saudi Arabia, Singapore, South Africa, Spain, Sri Lanka, United Arab Emirates, and the United States.

A.7 PriceWaterhouseCoopers

- Company website: www.pwc.com
- Careers: www.pwc.com/careers
- Facebook: www.facebook.com/pwc
- LinkedIn: www.linkedin.com/company/pwc
- Twitter: twitter.com/pwc

PriceWaterhouseCoopers is a multinational professional services network of firms (separate legal entities), operating under the PwC brand. "PwC" is often used to refer either to individual firms within the PwC network or to several or all of them collectively.

Actuarial careers at PwC

PwC's more than 320,000 employees provide services to address clients' needs in risk management, customer engagement, operational improvement, finance intelligence, and workforce experience. For clients in the insurance industry, more than 1400 insurance-dedicated accountants, actuaries, consultants, and deals specialists have extensive experience providing thoughtful and practical insights to life, P&C, specialty insurers and reinsurers on practically every issue the industry faces.

PwC offers insurance clients consulting services in various aspects of insurance business, including actuarial and financial modernization, captive insurance, cloud computing, risk modeling, insurance business strategy, insurance tax, policy, claims and billing transformation, and risk and capital management.

Actuarial Services in June 2023 had multiple openings for multiple titles in various locations of the United States. The titles included health insurance actuary manager, health insurance actuary senior associate and associate, actuarial life service – modernization director, actuarial life service—Prophet director, risk modeling service—P&C manager and associate, risk modeling service—life manager and experienced associate, climate scientist/climate risk modeling.

A.8   Willis Tower Watson

- Company website: www.wtwco.com
- Careers: careers.wtwco.com/actuarial
- Facebook: www.facebook.com/WTWcorporate
- LinkedIn: www.linkedin.com/company/wtwcorporate/
- Twitter: twitter.com/WTWcorporate

Willis Towers Watson Public Limited Company (Nasdaq: WTW) operates as an advisory, broking, and solutions company worldwide. It operates through two segments: health, wealth, and career; and risk and broking. The company offers actuarial support, plan design, and administrative services for traditional pension and retirement savings plans; plan management consulting, broking, and administration services for health and group benefit programs; and benefits outsourcing services. It also provides advice, data, software, and products to address clients' total rewards and talent issues. In addition, the company offers risk advice, insurance brokerage, and consulting services in the areas of property and casualty, aerospace, construction, and marine. Further, it offers investment consulting and discretionary management services to insurance and reinsurance companies, insurance consulting and technology, risk and capital management, pricing and predictive modeling, financial and regulatory reporting, financial and capital modeling, merger and acquisition, outsourcing, and business management services; wholesale insurance broking services to retail and wholesale brokers; and underwriting and capital management, capital market, and advisory and brokerage services.

Actuarial careers at WTW

Actuaries at Willis Towers Watson consult on financial risk, and work on benefit plan pricing, funding, reserve calculations, and benchmarking studies on the clients' behalf. Actuaries have opportunities across a number of different business areas. Willis Towers Watson has more than 150 years' experience in actuarial consulting—from introducing the stable value plan in the United States to designing the first flexible pension plan in Canada to developing the first ever large-scale retiree bulk lump sum program in the United States.

The search for the keyword "actuarial" on the company's career webpage on June 8, 2023 returned 151 entry-level openings and junior and senior positions in Bermuda, Canada, Czech Republic, Denmark, France, Great Britain, Germany, Hong Kong, India, Ireland, Malaysia, Mexico, Netherlands, Philippines, Portugal, Saudi Arabia, Singapore, Spain, South Africa, Switzerland, Thailand, and the United States.

Appendix B

Insurance companies

AM Best, a global credit agency, news publisher, and data analytics provider specializing in the insurance industry, published in February 2023 the World's Largest Insurers—2023 Edition on its publication *Best's Review*. There are two separate sets of ranking, each with the top 25 insurance companies measured by nonbanking assets, and by net premium written for the fiscal year 2021. Ten companies appear on both lists:

Allianz SE (Germany)
Assicurazioni Generali S.p.A (Italy)
AXA S.A. (France)
Berkshire Hathaway Inc (US)
China Life Insurance (Group) Company (China)
Dai-ichi Life Holdings, Inc (Japan)
Life Insurance Corporation of India (India)
Nippon Life Insurance Company (Japan)
Ping An Ins (Grp) Co of China Ltd. (China)
Zurich Ins Grp Ltd (Switzerland)

These companies represent life insurance, property-casualty insurance, and reinsurance industries across Europe and Asia. All of these companies are briefly profiled here. All of these companies, through their insurance subsidiaries, employ actuaries.

The descriptions that follow are merely meant to be starting points for your career search. You may want to dig deeper into their subsidiaries' websites for detailed job opening information. As you browse through them, you should get a sense of what it is like to work as an actuary for different kinds of insurance companies. For this reason, the individual profiles stress different aspects of employment: global mobility, daily tasks, required qualifications, company philosophy, working conditions, and so on. The quoted material can be found on the websites of the respective companies. You should consult these sites for more complete information.

Actuaries' Survival Guide, Third Edition. DOI: 10.1016/B978-0-443-15497-3.00010-3

B.1 Allianz Group

- Website: www.allianz.com
- Careers: https://careers.allianz.com/
- Facebook: www.facebook.com/allianz
- LinkedIn: www.linkedin.com/company/allianz

Allianz is a German multinational financial services company headquartered in Munich, Germany. The company is a component of the DAX, Euro Stoxx 50, and Stoxx Europe 600 Insurance stock market indices. The Allianz Group is one of the world's leading insurers and asset managers. With around 159,000 employees worldwide, the Allianz Group serves 122 million customers in more than 70 countries. Allianz customers benefit from a broad range of personal and corporate insurance services, ranging from property, life, and health insurance to assistance services to credit insurance and global business insurance. On the insurance side, Allianz is the market leader in the German property-casualty and life insurance markets and has a strong international presence. Allianz Group conducts business in almost every European country, with Germany, Italy, and France being its most important markets. In addition, the Group has divisions in the United States, Central and Eastern Europe, and the Asia-Pacific region. In fiscal year 2022, the Allianz Group achieved total revenues of approximately 153 billion euros. Allianz is one of the world's largest asset managers, with third-party assets of 1635 billion euros under management at year end 2022.

B.2 Assicurazioni Generali S.p.A

- Website: www.generali.com
- Careers: www.generali.com/work-with-us/join-us/job-search
- Facebook: www.facebook.com/allianz
- LinkedIn: www.linkedin.com/company/generali/jobs/

Generali (Assicurazioni Generali) is one of the largest global insurance and asset management providers. Established in 1831 in Italy, it is present in over 50 countries in the world, with a total premium income of €81.5 billion in 2022. With 82,000 employees serving 68 million customers, the Group is the market leader in Italy, has a leading position in Europe, and a growing presence in Asia and Latin America.

At the heart of Generali's strategy is its Lifetime Partner commitment to customers, achieved through innovative and personalized solutions, best-in-class customer experience, and its digitalized global distribution capabilities. The Group has fully embedded sustainability into all strategic choices, with the aim to create value for all stakeholders while building a fairer and more resilient society.

B.3   AXA

- Website: www.axa.com
- Careers: www.axa.com/en/careers
- Facebook: www.facebook.com/axa
- LinkedIn: www.linkedin.com/company/axa

AXA's origins can be traced back to a small mutual insurer from Normandy, France. The name AXA first appeared in the 1980s after several strategic mergers and acquisitions. Since then, the Group has expanded and grown into an international leader in insurance and asset management. Today AXA is present in geographically diverse markets, with operations concentrated in Europe, North America, and Asia Pacific.

Protection has always been AXA's core business. Through its four business lines—property and casualty insurance; health coverage; life and savings products; and asset management—the Group helps individuals, companies, and societies to thrive. As of end of 2022, AXA was the global leader in commercial lines insurance, the second largest European insurer, and an international leader in employee benefits, serving 93 million clients in 51 countries through 145,000 employees and agents across the world.

B.4   Berkshire Hathaway Inc

- Website: www.berkshirehathaway.com
- LinkedIn: www.linkedin.com/company/berkshire-hathaway/

Berkshire Hathaway Inc. is an American multinational conglomerate holding company headquartered in Omaha, Nebraska. Its main business and source of capital is insurance. The company's insurance brands include auto insurance company GEICO, reinsurance company General Re, and Allegany Corporation, a provider in the reinsurance, excess and surplus, and specialty insurance markets. Its noninsurance subsidiaries operate in diverse sectors such as confectionery, retail, home furnishing, jewelry, apparel, electrical power, and natural gas distribution.

Berkshire is one of the largest components of the S&P 500 index and the top-ranked company in the Forbes Global 2000.

Berkshire's insurance and reinsurance activities are conducted through approximately 70 domestic and foreign-based insurance companies. Domestic insurance businesses focus primarily on property and casualty risks. In addition, through the operation of General Re, Berkshire's insurance business entities also include life, accident, and health reinsurers, as well as internationally based property and casualty reinsurers. Berkshire's insurance companies maintain capital strength at exceptionally high levels.

B.5 China Life Insurance (Group) Company

- Website: www.chinalife.com.cn

Insurance industry in China since 1949 had been monopolized by the People's Insurance Company of China (PICC) until late 1980s. The Insurance Law adopted in 1995 requires separation of life insurance and property insurance. The independent operation of life insurance business stemming from the PICC was established in 1996 and renamed China Life Insurance Company in 1999. Four years later, the company was reorganized and restructured into China Life Insurance (Group) company. The Group's wholly owned subsidiary, China Life Insurance Company. Ltd (stocks traded on Hong Kong Stock Exchange and the Shanghai Stock Exchange) is now China's largest life insurance provider and one of the Fortune Global 500 companies.

B.6 Dai-ichi Life Holdings, Inc

- Website: www.dai-ichi-life-hd.com
- LinkedIn: www.linkedin.com/company/dai-ichi-life-hd/

Japan's first mutual insurance company, Dai-ichi Life, was established in 1902, and eventually developed to Dai-ichi Life Holdings, Inc. In 1975, Dai-ichi Life established its first overseas representative office in New York to study U.S. insurance, economic, and financial systems as well as to promote international group insurance policies among local subsidiaries of Japanese corporations. Dai-ichi Life entered into the investment trust business in 1998, then demutualized and became a stock company in 2010. In 2017 Dai-ichi Life shifted to a Holding Company Structure. As of 2023, the Group is present in nine countries, holds total assets of 63.4 trillion Japanese Yen, and serves 11.53 million customers in Japan and more overseas through 62,260 employees, of which 13.3% are based abroad.

B.7 Life Insurance Corporation of India

- Website: www.licindia.in
- LinkedIn: www.linkedin.com/company/lic/about/

The Life Insurance Corporation of India is the largest insurance company in India, established in 1956 after the Parliament of India passed the Life Insurance of India Act. It was formed by merging 245 insurance companies and provident societies, whether Indian or foreign, that were operating in the country at the time, to complete nationalization of the life insurance sector in order to make insurance service accessible to a broader population. In the fiscal year ended March 31, 2023, the LIC sold 20.4 million policies, reported an increase of 16.67% in first-year premiums and dominated the country's life insurance market with a share of 62.58%.

B.8   Nippon Life Insurance Company

- Website: www.nissay.co.jp/english

Nippon Life was founded in July 1889 at the time when a premium table based on unique Japanese mortality experience was created. Nippon Life was the first Japanese life insurer to offer profit dividends to policyholders, which embodied the spirit of mutual aid. Since the end of World War II, the company, reborn as Nippon Life Insurance Company in 1947, persistently strives to embody the philosophy of "co-existence, co-prosperity, and mutualism" as a mutual company. Today Nippon Life is the leading life insurance company in Japan and one of the largest mutual life insurance companies in the world. The company operates life insurance business in seven countries outside of Japan, including the United States, Australia, India, Myanmar, China, Thailand, and Indonesia.

B.9   Ping An Ins (Group) Company of China Ltd

- Website: group.pingan.com
- LinkedIn: www.linkedin.com/company/ping-an

Ping An (meaning "Peace and Safety") Insurance Company was founded in 1988 to offer property, cargo freight, and liability insurance in an effort to meet the rising demand from businesses of which the number increased at exponential rate in the new era of Economic Reform and Opening. In 1994 the company launched its first life insurance product. Ten years later, the stocks of Ping An began trading on the Hong Kong Stock Exchange; the IPO raised HKD14.3 billion, making it the largest one in Hong Kong's history at the time. In 2007, Ping An Group was listed on the Shanghai Stock Exchange, as the world's largest IPO of an insurance company at that time.

Ping An Group is a Fortune Global 500 company that engages in integrated finance and healthcare business through its subsidiaries, including Ping An Life, Ping An Property & Casualty, Ping An Annuity, Ping An Health Insurance, Ping An Bank, Ping An Trust, Ping An Securities, Ping An Fund Management and Ping An Health, and Ping An Insurance Overseas (Holdings) Ltd, serving more than 227 million customers and more than 693 million internet users of its ecosystems in healthcare, financial services, auto services and smart city services.

B.10   Zurich Insurance Group Ltd

- Website: www.zurich.com
- Careers: www.zurich.com/careers/careers-at-zurich
- LinkedIn: www.linkedin.com/company/zurich-insurance-company-ltd

Zurich Insurance Group, founded in 1872 and headquartered in Zurich, Switzerland, provides insurance and financial services in more than 215 countries and territories. The Group's property-casualty insurance products include motor, property, liability, marine, and specialty lines coverage. In addition, the Group's life insurance, savings, and investment solutions for individuals and corporations focus on protection, retirement planning, and wealth management. In the United States, it operates the Farmers Insurance Group to offer a wide range of personal and commercial insurance products. The company was named one of the World's Most Admired Companies by Fortune Magazine in 2021.

Appendix C

Reciprocity agreements

The skillsets expected of actuaries are highly consistent from country to country, except for the knowledge of country-specific insurance laws and regulations. To make actuaries' accreditation status "portable" to facilitate relocation of actuaries between countries, actuarial organizations enter agreements to allow one association's fellow members to be recognized as fellows of another association without much extra (mostly repetitive) efforts by actuaries who seek relocating across country borders. Such agreements are referred to as *reciprocity agreements* or *mutual recognition agreements*.

The agreement describes the process, country by country, by which actuaries in the countries involved can become members of the actuarial societies in the other participating countries. The document is the official description of the reciprocity agreement and should be consulted for specific details. The key clauses of the agreement between the Institute and Faculty of Actuaries and the Institute of Actuaries of Australia are summarized below for illustration.

The agreement says, in essence, that actuaries who have become Fellows of a national actuarial society by the normal route (having passed the necessary examinations), and who are members in good standing (having paid the annual membership fee in their home country), meet the professionalism requirements of the guest country, fulfill the necessary residency requirements, and intend to practice in the guest country, can do so by reciprocity.

C.1 Mutual Recognition Agreement Example: MRA Between Institute and Faculty of Actuaries and the Institute of Actuaries of Australia

Fellows of the Institute of Actuaries of Australia in good standing can become Fellows of the Institute and Faculty of Actuaries on the following conditions:

- They have attained Fellowship of the IAAust by completing the qualification requirements, and not solely in recognition of membership of another actuarial association.

Actuaries' Survival Guide, Third Edition. DOI: 10.1016/B978-0-443-15497-3.00011-5
229

- They are entitled to practice as a member of the IAAust.
- They have at least 3 years' postqualification practical work-based experience in actuarial practice.
- They undertake to adhere to such professionalism requirements as are required of IFoA Fellows from time to time; and
- At the same time as applying, they authorize in writing the IAAust to release relevant records to the IFoA concerning any adverse disciplinary determination, finding, sanction and/or penalty, to which the applicant has been subject, in accordance with the IAAust's disciplinary process. Such records may be taken into consideration by the IFoA in considering the application, and may be retained by the IFoA thereafter for as long as is reasonably necessary.

Conversely, the Institute of Actuaries of Australia will admit to Fellowship of the IAAust a Fellow of the Institute and Faculty of Actuaries who wishes to pursue actively the profession of actuary in Australia, provided that they satisfy the following conditions:

- They have qualified as Fellows of the Institute and Faculty of Actuaries through examination, and not solely in recognition of membership of another actuarial association.
- They are entitled to practice as a member of the IFoA.
- They have at least 3 years' postqualification practical work-based experience in actuarial practice.
- They undertake to complete a professionalism course within 12 months of admission; and
- At the same time as applying, they authorize in writing the IFoA to release relevant records to the IAAust concerning any adverse disciplinary determination, finding, sanction and/or penalty to which the applicant has been subject, in accordance with the IFoA's disciplinary scheme. Such records may be taken into consideration by the IAAust in considering the application, and may be retained thereafter by the IAAust for as long as is reasonably necessary.

Anyone interested in taking advantage of this agreement should consult the relevant documents published on the Institute and Faculty of Actuaries website: actuaries.org.uk/membership/mutual-recognition.

C.2 Organizations That Enter Mutual Recognition Agreements

The following table presents a set of organizations (host) and counter associations (guests) with whom the host maintains MRAs that give fellows of the host the fellowship of the guest.

Host	Guests
Actuarial Association of Europe (AAE)	Full members of AAE (see C.3)
Canadian Institute of Actuaries	Actuarial Society of South Africa Institute of Actuaries of Australia Society of Actuaries in Ireland
Casualty Actuarial Society	Actuarial Society of South Africa Institute of Actuaries of Australia Institute of Actuaries of India
Institute of Actuaries of Australia	Actuarial Society of South Africa Canadian Institute of Actuaries Casualty Actuarial Society Institute of Actuaries of India Institute of Actuaries of Japan Institute and Faculty of Actuaries New Zealand Society of Actuaries Society of Actuaries Society of Actuaries in Ireland
Institute and Faculty of Actuaries*	Actuarial Society of South Africa Institute of Actuaries of Australia Institute of Actuaries of India Israel Association of Actuaries
Society of Actuaries	Institute of Actuaries of Australia Institute and Faculty of Actuaries Society of Actuaries in Ireland

*Following the introduction of Curriculum 2019, the IFoA reviewed all MRAs. As a non-EU/EEA member of the AAE, the IFoA has withdrawn from participation in the AAE's MRA as of October 15, 2021. No further updates can be found on the websites of IFoA and AAE.

C.3   MRAs Within the Actuarial Association of Europe

As of October 1, 2021, the Actuarial Association of Europe maintains reciprocity agreements for the recognition of actuarial qualification between the following national associations of actuaries in Europe:

Austria: The Austrian Actuarial Society
Belgium: Royal Association of Belgian Actuaries
Bulgaria: Bulgarian Actuarial Society
Channel Islands: Channel Islands Actuarial Association
Croatia: Croatian Actuarial Society
Cyprus: Cyprus Association of Actuaries
Czech Republic: Czech Society of Actuaries
Denmark: Danish Society of Actuaries
Estonia: Estonian Actuarial Society
Finland: Actuarial Society of Finland

France: Institute of Actuaries of France
Germany: German Actuarial Society
Greece: Hellenic Actuarial Society
Hungary: Hungarian Actuarial Society
Iceland: Society of Icelandic Actuaries
Ireland: Society of Actuaries in Ireland
Italy: National Council of Actuaries, the Italian Institute of Actuaries
Latvia: Latvian Actuarial Association
Lithuania: Lithuanian Actuarial Society
Luxembourg: Luxembourg Actuarial Association
Netherlands: Actuarial Society of the Netherlands
Norway: Norwegian Society of Actuaries
Poland: Polish Society of Actuaries
Portugal: Portuguese Institute of Actuaries
Romania: Romanian Association of Actuaries
Slovakia: Slovak Society of Actuaries
Slovenia: Slovenian Association of Actuaries
Spain: Spanish Institute of Actuaries, the Catalan Actuarial Association
Sweden: Swedish Society of Actuaries
Switzerland: Swiss Association of Actuaries
Turkey: Association of Actuaries of Turkey
United Kingdom: Institute and Faculty of Actuaries

Essentially, the agreement means that full-fledged actuaries have professional mobility throughout Europe. In addition, the following European societies have observer status: Malta Actuarial Society, Montenegrin Actuarial Society (Montenegro), Association of Actuaries of Serbia, and the Society of Actuaries of Ukraine.

Appendix D

Actuarial websites

D.1 Actuarial Organizations

1. Actuarial Association of Europe https://www.actuary.eu/
2. (All AAE member associations' websites can be found wherein)
3. Actuarial Foundations (www.actuarialfoundation.org)
4. Actuaries Institute (actuaries.asn.au)
5. Actuarial Society of Hong Kong (www.actuaries.org.hk)
6. Actuarial Society of South Africa (www.actuarialsociety.org.za)
7. American Academy of Actuaries (www.actuary.org)
8. American Society of Pension Professionals and Actuaries (www.asppa.org)
9. Canadian Institute of Actuaries (www.cia-ica.ca)
10. Casualty Actuarial Society (www.casact.org)
11. China Association of Actuaries (e-caa.org.cn)
12. Conference of Consulting Actuaries (www.ccactuaries.org)
13. Institute of Actuaries of India (actuariesindia.org)
14. Institute of Actuaries of Japan (www.actuaries.jp/english)
15. Institute and Faculty of Actuaries (actuaries.org.hk)
16. Instituto Brasileiro de Atuária (Brazil) (atuarios.org.br)
17. International Actuarial Association (https://www.actuaries.org/iaa)
18. Society of Actuaries (www.soa.org)

D.2 Actuarial Publishers and Exam Preparation Services

1. ACTEX Learning/Mad River Books (www.actexmadriver.com)
2. ACTEX Learning (www.actexlearning.com)
3. Actuarial Bookstore (www.actuarialbookstore.com)
4. Actuarial Study Materials (www.studymanuals.com)
5. Be an Actuary (www.beanactuary.org)
6. Coaching Actuaries (www.coachingactuaries.com)
7. The Infinite Actuary (https://www.theinfiniteactuary.com/)

Actuaries' Survival Guide, Third Edition. DOI: 10.1016/B978-0-443-15497-3.00012-7

D.3   Test Centers

1. Pearson VUE Testing Centers for Casualty Actuarial Society (home.pearsonvue.com/cas)
2. Prometric Center for Society of Actuaries (www.prometric.com/soa)

Appendix E

Actuarial recruiting agencies

E.1 Prominent Agencies

The following list of actuarial recruiting agencies form a small set of the large number of well-known companies in the United States, Canada, the UK, and elsewhere that can serve as a starting point for you in your search for an ideal actuarial position worldwide. You can get information on the status of these companies and others by consulting Dun and Bradstreet at www.dnb.com, the world's leading source of commercial information and insight into businesses. In addition, you will find job listings using actuary-specific search engines such as https://www.actuary.com/. General career service companies are not profiled here.

- Actuarial Careers, Inc. (White Plains, NY) (www.actuarialcareers.com). Exclusively dedicated to the placement and advancement of Chief Actuaries, Fellows, Associates, and Student Actuaries.
- Acumen Resources (London, UK) (www.acumen-resources.com). Acumen Resources is one of the world's leading actuarial recruitment agencies. All the consultants have either actuarial or human resources qualifications and have worked in a wide variety of practice areas. This allows clients to tap into a significant global network and maximize their chances of finding the right actuarial job with the right company.
- Andover Research (New York, NY) (www.andoverresearch.com). Andover Research, Ltd. specializes in the recruitment and placement of actuaries and investment professionals since 1975. The firm's actuarial recruiters have the experience, expertise, and insight to direct clients to their next opportunity.
- DW Simpson and Company (Chicago, IL) (www.dwsimpson.com). DW Simpson Global Actuarial Recruitment serves the actuarial profession worldwide in all disciplines, recruiting at all levels from entry-level through fellowship, and works with clients on both retained and contingent searches. The company recruits actuaries for employment in casualty, health, life, and pension for Canada, the United States, the UK, Europe, India, Asia, Australia, the Middle East, and other regions.

Actuaries' Survival Guide, Third Edition. DOI: 10.1016/B978-0-443-15497-3.00013-9

- Darwin Rhodes (London, UK) (www.darwinrhodes.com). Darwin Rhodes has its headquarters in the City of London with specialist teams covering the UK market. The company covers the actuarial and insurance markets of mainland Europe and operates from offices in both London and Zurich, Switzerland. Its offices in Hong Kong and Shanghai cover the major financial centers of East and Southeast Asia including Tokyo and Singapore and emerging centers such as Kuala Lumpur, Taipei, Bangkok, and Ho Chi Minh City.
- Darwin Rhodes was the first insurance recruiter to enter the Indian market. Its New York and Chicago offices cover the major North American financial centers and developing markets in Central and South America. Its Sydney office covers both Australia and New Zealand.
- Elliot Bauer (London, UK) (www.elliottbauer.com). Since Elliott Bauer's inception, actuarial has always been a key specialism and the firm continues to assist clients across the globe while nurturing and guiding future talent in this field.
- Emerald Group (London, UK) (www.emerald-group.com). The Emerald Group originated in the UK as an organization serving the actuarial profession at all levels. It provides staff to many of the world's largest consultancy and investment houses across five continents.
- Ezra Penland Actuarial Recruitment (Chicago, IL) (www.ezrapenland.com). Ezra Penland is ingrained in the actuarial profession and has strong knowledge, experience, and contacts within it. The company is also an established presence within the predictive analytics, catastrophe modeling, and financial modeling disciplines.
- Mid America Search (West Des Moines, IA) (www.midamericasearch.com). Mid America Search is an executive search firm specializing primarily in the insurance and financial industry. It conducts personnel searches for hundreds of outstanding insurance companies and related organizations both in the United States and internationally.
- Pinnacle Group (Portsmouth, NH) (https://support40328.wixsite.com/pinnaclejobs). The Pinnacle Group specializes in actuarial recruitment. Its clients include leading insurance, consulting, and investment firms in the United States and elsewhere. Its areas of recruitment include life and annuity, health and managed care, property and casualty, pension and benefits, and reinsurance.
- Pryor Associates (Hicksville, NY) (www.ppryor.com). Pryor Associates agency provides employees for all facets of the insurance industry including administration, underwriting, claims, loss control, accounting, actuarial, pension, property and casualty, life and health, employee benefits, agency and brokerage, and insurance and reinsurance company placements.
- S.C. International (Downers Grove, IL) (www.scinternational.com). S.C. International, Ltd. focuses primarily on staffing of the actuarial, employee

benefits, and insurance market. Like many of the actuarial recruitment agencies, the company plays an integral role in both sourcing personnel and coordinating the interview process, from initial conversations to meetings and final negotiations.

- Smith Stanley Associates, LLC. (Southport, CT) (www.smithhanley.com). The mission of Smith Stanley Associates, LLC. is to concentrate on select areas including the actuarial profession to help clients reach their long-term success.

E.2   Sample Job Postings

Actuarial job postings can be found on the websites of employers, recruiting agencies, and actuarial organizations. You can simply enter the keywords like "actuarial" or "enter-level actuary" to get started, or search by job function like "reserving actuary" or "valuation actuary." The sample job postings presented here are from the career webpages of SOA (jobs.soa.org) and CAS (www.cas-act.org).

Annuity actuary

A search for "annuity actuary" produced 12 hits on the career page of the SOA website. Here is a sample:

Position Title:	Associate Actuary—Product Management
Location:	Nebraska
Workplace Type:	Hybrid
Area of Practice:	Insurance, Life
Job Type:	Full Time

Duties:
Provide support for developing life, health, and fixed annuity products, specifically:

- Determining pricing assumptions and building pricing models.
- Performing profit tests and sensitivity tests on pricing models.
- Preparing actuarial documents for product filings and responding to filing objectives.
- Providing oversight for product administration set-up, testing, and implementation.

Requirements:

- 3 to 5 years of experience in life insurance pricing.
- Bachelor's degree in actuarial science or related field.
- ASA designation.
- Good interpersonal, verbal, and written communication skills.

- Ability to interact and relate well with people in a wide variety of areas, including sales agents, state insurance departments, and all departments in the home office.
- Good personal computer skills including word processing and spreadsheet capabilities.

Life actuary

A search for "life actuary" produced 14 hits on the career page of the SOA website. Here is a sample:

Position Title:	AVP, Actuarial Financial Reporting
Location:	Maryland
Workplace Type:	Onsite
Area of Practice:	Financial Services—Investment Banking and Fund Management
Job Type:	Full Time

Duties:

- Supervise the data collection, organization, and reporting of actuarial results for financial statements.
- Lead the actuarial efforts in audits and coordinate efforts to address auditors' concerns.
- Lead the studies of mortality and lapse experience.
- Manage and maintain in-force projection models.

Requirements:

- Ten or more years in the life insurance industry.
- Bachelor's degree in mathematically-oriented major.
- F.S.A. credential.
- Extensive work with and understanding of valuation, forecasting, and product profitability analyses.
- Deep understanding of the U.S. Statutory reserving framework.
- Ability to build, communicate, and monitor team's development plans effectively.

Pricing actuary

A search for "pricing actuary" resulted in 16 hits on the career page of the CAS website. Here is a sample.

Position Title:	Associate Actuary
Location:	Tennessee
Type:	Full Time
Categories:	Predictive modeling, property-casualty, ratemaking

Duties:

- Manage actuarial rate indications and provide actuarial support to company rate filings.
- Create advanced pricing models, such as GLM, using statistical software.
- Prepare special data studies and reports.
- Oversee company compliance with statistical reporting requirements.
- Work collaboratively with Actuarial, Product Management, and Business Intelligence Departments on company projects.
- Train and mentor actuarial staff on methods and company practices.

Requirements:

- Education: BA or BS degree in Mathematics, Statistics, Actuarial Science, or related field.
- Credential: ACAS or FCAS credential.
- Experience: At least 4 years of insurance experience.
- Strong EXCEL skills.
- Experience with SQL, SAS, or R programming is preferred.
- Experience with creating GLM-based pricing models is preferred.

Reserving actuary

A search for "reserving actuary" produced eight hits on the career page of the CAS website. Here is a sample:

Position Title:	Actuarial Consultant
Location:	Nationwide
Type:	Full Time
Category:	Consulting, property/casualty, ratemaking
Additional Information:	Telecommuting is allowed

Duties:

- Ratemaking and reserving projects for both commercial and personal lines of business.
- Rate level indications, class plan analyses, benchmarking analyses, or competitor rate comparisons.
- Reserve analyses, self-insured retention analyses.
- Funding studies, pro formas, and more.
- Communicate and work directly with clients as needed, with supervision.

Requirements:

- 1 to 3 years' experience in the insurance industry.
- At least one completed Actuarial Exam.
- Knowledge of ratemaking of personal auto and homeowners lines of business.

- Excellent skills in Microsoft Excel, Word, and SQL.
- Good written and oral communication skills.

Risk modeler

A search for "risk modeler" resulted in 11 hits on the career page of the CAS website. Here is a sample:

Position Title:	Senior Actuarial Analyst—Product
Location:	Nationwide
Type:	Full Time
Category:	Property/casualty
Additional Information:	Telecommuting is allowed

Duties:

- Develop and test new rating approaches and rating variables.
- Evaluate rate adequacy and work with underwriters to ensure rates maintain adequacy.
- Monitor new business and renewal rate levels.
- Understand and explain external trends affecting each product.

Requirements:

- Bachelor's degree in Science, Technology, Engineering, or Math.
- 1 to 3 years of pricing experience in construction liabilities.
- Python and/or R programming knowledge, database and SQL programming knowledge.
- Excellent critical thinking, problem-solving, and communication skills.

Valuation actuary

A search for "valuation actuary" produced six hits on the career page of the SOA website. Here is a sample:

Position Title:	Entry-Level Actuarial Analyst—Pension
Location:	Georgia, Texas, Illinois, Florida
Workplace Type:	Hybrid
Area of Practice:	Consulting
Job Type:	Full Time

Duties:
Assist in preparation of recurring actuarial valuation and compliance work for qualified retirement plans of assigned clients, including:

- Actuarial funding valuations.
- Accounting valuations.

- Filings with insurance regulators.
- Providing input to actuary and consultant on appropriate steps.

Requirements:

- 0 to 2 years of experience.
- Bachelor's degree.
- Minimal of one actuarial examination (more is a plus).
- Strong oral and written communication skills.
- Pursuing actuarial exams with goal of becoming an ASA or FSA.
- Strong team player.

Appendix F

SOA education summary

F.1 Associateship Requirements

Validation by educational experience

- VEE Accounting and Finance
- VEE Economics
- VEE Mathematical Statistics

E-learning courses

- Preactuarial Foundations Module
- Actuarial Foundations Module
- Fundamentals of Actuarial Practice (FAP) e-learning course
- ATPA Assessment—Advanced Topics in Predictive Analytics

Examinations

- Exam P: Probability
- Exam FM: Financial Mathematics
- Exam FAM: Fundamentals of Actuarial Mathematics
- Exam SRM: Statistics for Risk Modeling
- Exam ALTAM or ASTAM (pick one):
    - Advanced Long-Term Actuarial Mathematics, or
    - Advanced Short-Term Actuarial Mathematics
- Exam PA: Predictive Analytics

Seminar

Associateship Professionalism Course (APC)

Actuaries' Survival Guide, Third Edition. DOI: 10.1016/B978-0-443-15497-3.00014-0

F.2   Fellowship Requirements

In addition to the Associateship requirements, a candidate for Fellowship must pass the Decision Making and Communication (DMAC) Module and the Fellowship Admissions Course (FAC), and must complete the modules and exams in one of the following six specialty tracks.

1. Corporate Finance and ERM Track

    - ERM Module
    - ERM Exam
    - Introduction to Corporate Finance and ERM Module
    - Foundations of CFE exam
    - Strategic Decision Making Exam
    - Advanced Topics in CFE Module

2. Quantitative Finance and Investment Track

    - ERM Module
    - Financial Modeling Module
    - Scenario Modeling Module
    - Quantitative Finance Exam
    - Portfolio Management Exam
    - Investment Risk Management Exam or ERM Exam

3. Individual Life Insurance and Annuities Track

    - Introduction to ILA Module
    - Regulation and Taxation Module
    - ERM Module
    - Life Product Management Exam
    - Life Financial Management Exam
    - Life ALM and Modeling Exam or ERM Exam

4. Retirement Benefits Track

    - Social Insurance Module
    - Funding and Regulation Exam (Canada only) or Enrolled Actuaries Exams 2 L and 2 F (United States only)
    - Design and Accounting Exam (United States or Canada)
    - ERM Module
    - Retirement Plan Investment and Risk Management Exam or ERM Exam
    - Pension Projections Module

5. Group and Health Track

    - Health Economics Module
    - Health Foundations Module
    - Design and Pricing Exam

- Finance and Valuation Exam
- Group and Health Specialty Exam or ERM Exam
- Pricing, Reserving and Forecasting Module or ERM Module

6. General Insurance Track

  - Introduction to General Insurance Exam
  - Ratemaking and Reserving Exam
  - Financial and Regulatory Environment Exam (United States or Canada)
  - Financial Economics Module
  - ERM Module
  - Applications of Statistical Techniques Module
  - Advanced Topics in General Insurance Exam or ERM Exam

F.3   Chartered Enterprise Risk Analyst (CERA) Requirements

Validation by educational experience

- VEE Accounting and Finance
- VEE Economics
- VEE Mathematical Statistics

E-learning courses

- Preactuarial Foundations Module
- Actuarial Foundations Module
- Enterprise Risk Management Module
- Fundamentals of Actuarial Practice (FAP) e-learning course

Examinations

- Exam P: Probability
- Exam FM: Financial Mathematics
- Exam FAM: Fundamentals of Actuarial Mathematics
- Exam SRM: Statistics for Risk Modeling
- Exam ERM: Enterprise Risk Management

Seminar

Associateship Professionalism Course (APC)

Appendix G

CAS education summary

G.1 Associateship Requirements

Validation by educational experience

- VEE Accounting and Finance
- VEE Economics

Online CAS data and insurance series courses

- Introduction to Data Analytics
- Risk Management and Insurance Operations (same as Course CA1 of the Institute of Actuaries (UK))
- Insurance Accounting, Coverage Analysis, Insurance Law and Insurance Regulation (same as Course CA2 of the Institute of Actuaries (UK))

Examinations

- Exam 1: Probability (same as SOA Exam P)
- Exam 2: Financial Mathematics (same as SOA Exam FM)
- Exam MAS I: Modern Actuarial Statistics I
- Exam MAS II: Modern Actuarial Statistics II
- Course on Professionalism
- Exam 5: Basic Techniques for Ratemaking and Estimating Claim Liabilities
- Exam 6: Regulation and Financial Reporting (Nation-Specific)

G.2 Fellowship Requirements

Examinations

- Exam 7: Estimation of Policy Liabilities, Insurance Company Valuation, and Enterprise Risk Management
- Exam 8: Advanced Ratemaking
- Exam 9: Financial Risk and Rate of Return

Actuaries' Survival Guide, Third Edition. DOI: 10.1016/B978-0-443-15497-3.00015-2
© 2025 Elsevier Inc. All rights reserved, including those for text and data mining, AI training, and similar technologies.

Appendix H

Actuarial symbols

True to the origin of their name, actuaries use an extensive list of special symbols for their work. It's a kind of cleverly devised shorthand for actuarial objects and functions. The notation is based on principles for construction adopted by the Second International Congress of Actuaries in London in 1898. The list is modified and updated from time to time with the approval of the Permanent Committee of Actuarial Notations of the International Actuarial Association. Appendices 3 and 4 of the standard reference on actuarial mathematics by Bowers et al. (1997) contain a full list of symbols. The following list gives an indication of the kind of symbols involved and is far from being exhaustive. Many of these symbols occur in the sample questions and answers in Chapter 2. For missing symbols used in these examples, we refer to Bowers et al. (1997) for their definitions and explanations.

- The symbol $a_{\overline{n}|}$ denotes the present value of an annuity that pays one dollar at the end of each year for n years.

- The symbol $\ddot{a}_{\overline{n}|}$ denotes the present value of an annuity that pays one dollar at the beginning of each year for n years.

- The symbol $a_{\overline{n}|i}$ denotes the present value of an annuity that pays one dollar at the end of each year for n years, evaluated at i percent interest per year.

- The symbol $\ddot{a}_{\overline{n}|i}$ denotes the present value of an annuity that pays one dollar at the beginning of each year for n years, evaluated at i percent interest per year.

- The symbol $\overline{a}$ denotes the present value of an annuity payable continuously at the rate of one dollar per year.

- The symbol (x) denotes a living person age x.

- The symbol (xy) denotes two living persons age x and y, respectively.

- The symbol $\ddot{a}_{xy}$ denotes the present value of an annuity payable during the joint lives of (x) and (y), payable at the beginning of each year.

Actuaries' Survival Guide, Third Edition. DOI: 10.1016/B978-0-443-15497-3.00016-4

- The symbol $\ddot{a}_{\overline{xy}}$ denotes the present value of an annuity payable as long as one of (x) and (y) is alive, payable at the beginning of each year.

- The symbol $\ddot{a}_{x:\overline{n}|}$ denotes the present value of an annuity payable, at the beginning of each year, during the joint duration of the life of (x) and a term of n years.

- The symbol A_x denotes the expected present value of a whole life insurance policy issued to (x) that pays 1 at the end of the year of death.

- The symbol $_nd_x$ denotes the expected number of deaths between the ages x and $x + n$.

- The symbol $\overset{\circ}{e}_x$ denotes the average remaining lifetime at age x.

- The symbol $(Ia)_{\overline{n}|i}$ denotes the present value of an annuity that pays one dollar at the end of year 1, two dollars at the end of year 2, ..., n dollars at the end of year n, at i percent interest per year.

- The symbol $(Da)_{\overline{n}|i}$ denotes the present value of an annuity that pays n dollars at the end of year 1, $n - 1$ dollars at the end of year 2, ...,1 dollar at the end of year n, at i percent interest per year.

- The symbol l_x denotes the expected number of survivors to age x from l_0 newborns.

- The symbol P_x denotes the probability that life (x) will reach age $x + 1$.

- The symbol q_x denotes the probability that life (x) will die within the next year.

- The symbol $_tP_x$ denotes the probability that life (x) will survive the next (t) years.

- The symbol $_tq_x$ denotes the probability that life (x) will die within the next (t) years.

- The symbol P_x denotes the annual premium of a whole life policy of one dollar, payable at the end of the year of the death of (x), with the first premium payable when the policy is issued.

- The symbol P_{xy} denotes the annual premium of a whole life policy of one dollar, payable at the end of the year of the first death of (x) or (y), with the first premium payable when the policy is issued.

- The symbol $s(x)$ denotes the probability that a newborn will reach age x.

- The symbol $s_{\overline{n}|}$ denotes the accumulated value of an annuity of one dollar per year for n years, payable at the end of each year.

Appendix I

Bibliography

Actuarial Association of Europe. (2019). *Core syllabus for actuarial training in Europe*. actuary.eu/wp-content/uploads/2019/10/2019-10-11_AAE-Core-Syllabus_complete_final.pdf

Barhoumeh, D., September 2018. Access to actuarial education in the Arab world. *International News*, Issue 75, The Society of Actuaries.

Best, A.M., December 2022. Top audit and actuarial firms. *Best's Review, 123*(12), 19–21.

Best, A.M., February 2023. World's largest insurers—2023 edition. *Best's Review, 124*(2), 42–44.

Bowers, N.L., Gerber, H.U., Hickman, J.C., Jones, D.A., Nesbitt, C.J., 1997. *Actuarial mathematics*, 2nd ed. The Society of Actuaries.

Brown, R.L., 2002. The globalization of actuarial education. *British Actuarial Journal, 8*(1), 1–3.

Bureau of Labor Statistics, U.S. Department of Labor. (2023). *Occupational outlook handbook*, Actuaries. https://www.bls.gov/ooh/math/actuaries.htm

Casualty Actuarial Society and Society of Actuaries. (2023). *Offering a strong combination of high salary and job security*. www.beanactuary.org/what-is-an-actuary/a-top-ranked-job

Dickson, C.M.D., Hardy, M.R., Waters, H.R., 2020. *Actuarial mathematics for life contingent risks*, 3rd ed. Cambridge University Press.

Institute and Faculty of Actuaries. (2011). *Actuaries in risk management*. www.actuaries.org.uk.

Institute and Faculty of Actuaries. (2017). *Risk management—An actuarial approach*. www.actuaries.org.uk

Actuaries' Survival Guide, Third Edition. DOI: 10.1016/B978-0-443-15497-3.00017-6

International Actuarial Associations. (2017). *2017 IAA education syllabus.* www.actuaries.org/CTTEES_EDUC?Documents/2017_IAA_Education_ Syllabus.pdf

Klugman, S., Panjer, H., Willmot, G., 2019. *Loss models: From data to decisions*, 5th ed. John Wiley & Sons, Inc.

Krantz, L., Lee, T., 2015. *Jobs rated Almanac—The best jobs and how to get them*, 7th ed. CreateSpace Independent Publishing Platform.

Perryman, F.S., 1949. International Actuarial Notation. *Proceedings of the Casualty Actuarial Society, 36*(66), 123–131.

Schwartz, L., Douglas, M., 2019. *Big data and the future actuary*. Society of Actuaries. www.soa.org/globalassets/assets/files/resources/research-report/ 2019/big-data-future-actuary.pdf

Index

Page numbers followed by "*f*" indicate figures and "*t*" indicate tables.